the
NOISY
PENDULUM

the
NOISY
PENDULUM

Moshe Gitterman

Bar-Ilan University, Israel

World Scientific

NEW JERSEY · LONDON · SINGAPORE · BEIJING · SHANGHAI · HONG KONG · TAIPEI · CHENNAI

Published by

World Scientific Publishing Co. Pte. Ltd.

5 Toh Tuck Link, Singapore 596224

USA office: 27 Warren Street, Suite 401-402, Hackensack, NJ 07601

UK office: 57 Shelton Street, Covent Garden, London WC2H 9HE

Library of Congress Cataloging-in-Publication Data
Gitterman, M.
  The noisy pendulum / Moshe Gitterman.
    p. cm.
  Includes bibliographical references and index.
  ISBN-13: 978-981-283-299-3 (hardcover : alk. paper)
  ISBN-10: 981-283-299-8 (hardcover : alk. paper)
  1. Pendulum. 2. Noise. 3. Mechanics. 4. Physics. I. Title.
  QA862.P4.G58 2008
  531'.324--dc22

                            2008032349

British Library Cataloguing-in-Publication Data
A catalogue record for this book is available from the British Library.

Printed in Singapore.

Preface

Although every modern house has a TV set and a computer, one can still find here and there, a grandfather wall clock going tick-tock. The invention of this clock goes back to 1581, when the 17 year-old Galileo Galilei watched a suspended lamp swinging back and forth in the cathedral of Pisa, and found, to his surprise, that it took the same number of pulse beats for the chandelier to complete one swing, no matter how large the amplitude. The larger the swing, the faster the motion, but the time was the same. Therefore, time could be measured by the swing of a pendulum – the basis for the pendulum clock. In 1602, Galilei explained the isochronism of long pendulums in a letter to a friend, and a year later, another friend, Santorio Santorino, a physician in Venice, began using a short pendulum, which he called a "pulsilogium", to measure the pulse of his patients. This discovery had important implications for the measurement of time intervals[1]. Following Galilei, a few other famous scientists continued the analysis of the pendulum. For example, the period of rotation and oscillation with finite amplitude were calculated by Huygens (1673) and by Euler (1736), respectively.

Although the legend of how Galilei discovered this property of the simple pendulum is probably apocryphal, the main idea of a pendulum is widely used today to describe the oscillation of prices in the stock market or the

---

[1] As a matter of fact, one of the earliest uses of the pendulum was in the seismometer device of the Han Dynasty (202 BC – 220 AD) by the scientist and inventor Zhang Heng (78 – 139). Its function was to sway and activate a series of levers after being disturbed by the tremor of an earthquake. After being triggered, a small ball would fall out of the urn-shaped device into a metal toad's mouth, signifying the cardinal direction where the earthquake was located (and where government aid and assistance should be swiftly sent). Also, an Arabian scholar, Ibn Vinus, is known to have described an early pendulum in the 10th century [1].

change of mood of our wives (husbands). However, scientists use the concept of a pendulum in a much more comprehensive sense, considering it as a model for a great diversity of phenomena in physics, chemistry, economics and communication theory.

A harmonic oscillator is the simplest linear model, but most, if not all physical processes are nonlinear, and the simplest model for their description is a pendulum. Tremendous effort has gone into the study of the pendulum. From the enormous literature it is worth mentioning the bibliographic article [2], the International Pendulum conference [3], and the recent book [4] written in a free, easy-going style, in which calculations alternate with historical remarks.

The aim of the present book is to give the "pendulum dictionary", including recent (up to 2007) results. It can be useful for a wide group of researchers working in different fields where the pendulum model is applicable. We hope that teachers and students will find some useful material in this book. No preliminary knowledge is assumed except for undergraduate courses in mechanics and differential equations. For the underdamped pendulum driven by a periodic force, a new phenomenon - deterministic chaos - comes into play, and one has to take into account the delicate balance between this chaos and the influence of noise.

The organization of the book is as follows.

Part 1 is comprised of three sections. Sec. 1.1 contains the description and solution of the dynamic equation of motion for the simple mathematical pendulum oscillating in the field of gravity, which creates a torque. This equation is isomorphic to a model description of different phenomena, which are presented in Sec. 1.2. The description of noise as used in the book is given in Sec. 1.3.

Part 2 describes the properties of the overdamped pendulum. The simple overdamped case for which one neglects the inertial term is considered first, thereby reducing the second-order differential equation to one of first-order. The analytical solution of the overdamped deterministic pendulum is described in Sec. 2.1. The addition of additive and multiplicative noise, both white and dichotomous, provide the subject of Sec. 2.2. An overdamped pendulum subject to a periodic torque signal is described in Sec. 2.3, in particular, the analytic solution for a periodic force having the form of a pulsed signal.

Part 3 is devoted to the underdamped pendulum, including the effect of friction (Sec. 3.1), multiplicative noise (Sec. 3.2), additive noise (Sec. 3.3), the periodically driven pendulum (Sec. 3.4), overall forces (Sec. 3.5), and an oscillating suspension point (Sec. 3.6). The spring pendulum is described in Sec. 3.7. Finally, resonance-type phenomena are considered in Sec. 3.8.

Part 4 is devoted to a description of the phenomenon of deterministic chaos. Deterministic chaos leads to "noise-like" solutions, which explains why this chapter appears in the noisy pendulum book. The general concepts are introduced in Sec. 4.1, and the transition to chaos for different cases provides the subject matter for Sec. 4.2. New phenomena (chaos control, erratic motion and vibrational resonance), which are caused by an addition of a second periodic force, are considered in Sec. 4.3.

Part 5 contains the analysis of the inverted pendulum. The inverted position can be stabilized either by periodic or random oscillations of the suspension axis or by the influence of a spring inserted into a rigid rod. These three cases are considered in Secs. 5.1–5.4, while the joint effect of these factors is described in Sec. 5.5. Finally, we present our conclusions.

Two comments are appropriate. First, in spite of the fact that the majority of results presented in this book are related to nonintegrable equations which demand numerical solutions, we are not discussing technical questions concerning the methods and accuracy of numerical calculations. Second, we consider only a single one-dimensional classical pendulum. The quantum pendulum and interactions between pendula are beyond the scope of this book. Even then, there are a tremendous number of published articles devoted to the pendulum (331 articles in the American Journal of Physics alone). In order not to increase unduly the size of this book and to keep it readable, I have not described all these articles. I ask the forgiveness of the authors whose publications remain beyond the scope of this book.

Contents

Chapter 1

Formulation of the Problem

1.1 Mathematical pendulum

The pendulum is modeled as a massless rod of length l with a point mass (bob) m of its end (Fig. 1.1). When the bob performs an angular deflection ϕ from the equilibrium downward position, the force of gravity mg provides a restoring torque $-mgl \sin \phi$. The rotational form of Newton's second law of motion states that this torque is equal to the product of the moment of inertia ml^2 times the angular acceleration $d^2\phi/dt^2$,

$$\frac{d^2\phi}{dt^2} + \frac{g}{l} \sin \phi = 0. \tag{1.1}$$

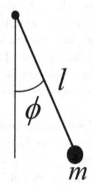

Fig. 1.1 Mathematical pendulum.

If one introduces damping, proportional to the angular velocity, Eq. (1.1) takes the following form:

$$\frac{d^2\phi}{dt^2} + \gamma\frac{d\phi}{dt} + \frac{g}{l}\sin\phi = 0. \tag{1.2}$$

Equations (1.1) and (1.2) are called the underdamped equations. In many cases, the first, inertial term in (1.2) is small compared with the second, damping term, and may be neglected. Then, redefining the variables, Eq. (1.2) reduces to the overdamped equation of the form

$$\frac{d\phi}{dt} = a - b\sin\phi. \tag{1.3}$$

For small angles, $\sin\phi \approx \phi$, and Eqs. (1.1)-(1.3) reduce to the equation of a harmonic oscillator. The influence of noise on an oscillator has been considered earlier [5].

Let us start with the analysis based on the $(\phi, d\phi/dt)$ phase plane. Multiplying both sides of Eq. (1.1) by $d\phi/dt$ and integrating, one obtains the general expression for the energy of the pendulum,

$$E = \frac{l^2}{2}\left(\frac{d\phi}{dt}\right)^2 + gl\left(1 - \cos\phi\right) \tag{1.4}$$

where the constants were chosen in such a way that the potential energy vanishes at the downward vertical position of the pendulum. Depending on the magnitude of the energy E, there are three different types of the phase trajectories in the $(\phi, d\phi/dt)$ plane (Fig. 1.2):

1. $E < 2gl$. The energy is less than the critical value $2gl$, which is the energy required for the bob to reach the upper position. Under these conditions, the angular velocity $d\phi/dt$ vanishes for some angle $\pm\phi_1$, i.e., the pendulum is trapped in one of the minima of the cosine potential well, performing simple oscillations ("librations") around the position of the minimum. This fixed point is called an "elliptic" fixed point, since nearby trajectories have the form of ellipses.

2. $E > 2gl$. For this case, there are no restrictions on the angle ϕ, and the pendulum swings through the vertical position $\phi = \pi$ and makes complete rotations. The second fixed point $(\pi, 0)$, which corresponds to the pendulum pointing upwards, is a "hyperbolic" fixed point since nearby trajectories take the form of a hyperbola.

3. $E = 2gl$. For this special case, the pendulum reaches the vertical position $\phi = \pi$ with zero kinetic energy, and it will remain in this unstable

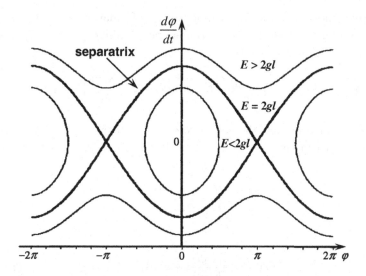

Fig. 1.2   Phase plane of a simple pendulum described by Eq. (1.1).

point until the slightest perturbation sends it into one of the two trajectories intersecting at this point. The border trajectory, which is located between rotations and librations, is called the separatrix, since it separates different types of motion (oscillations and rotations). The equation of the separatrix can be easily obtained from (1.4),

$$\frac{d\phi}{dt} = 2\sqrt{\frac{g}{l}} \cos \frac{\phi}{2}. \tag{1.5}$$

The time t needed to reach the angle ϕ is given by

$$t = \sqrt{\frac{l}{g}} \ln \left[\tan \left(\frac{\phi}{4} + \frac{\pi}{4} \right) \right]. \tag{1.6}$$

Trajectories close to the separatrix are very unstable and any small perturbations will result in running or locked trajectories. These trajectories possess interesting properties [6]. Two trajectories intimately close to the separatrix, with energies 0.9999 $(2gl)$ and 1.0001 $(2gl)$, describe the locked and running trajectories, respectively. In spite of having almost the same energy, their periods differ by a factor of two (!) so that the period of oscillation is exactly twice the period of rotation. Physical arguments support this result [6].

It is convenient to perform a canonical transformation from the variables ϕ and $d\phi/dt$ to the so-called action-angle variables J and Θ [7]. For librations, the action J is defined as

$$J = \frac{1}{2\pi} \oint \left(\frac{d\phi}{dt} \right) d\phi = \frac{\sqrt{2}}{l} \int_{-\phi_1}^{\phi_1} d\phi \sqrt{E - gl + gl \cos\phi} = \frac{8\sqrt{gl}}{\pi} [E(\kappa) - \kappa^2 K(\kappa)]$$

$$(1.7)$$

where $K(\kappa)$ and $E(\kappa)$ are the complete elliptic integrals of the first and second kind with modulus $\kappa = \sqrt{E/2gl}$. The angle Θ is defined by the equation

$$\frac{d\Theta}{dt} = \frac{\partial E}{\partial J} = \frac{\pi\sqrt{gl}}{2K(\kappa)}, \qquad (1.8)$$

yielding

$$\Theta(t) = \frac{\pi\sqrt{gl}}{2K(\kappa)} t + \Theta(0). \qquad (1.9)$$

One can easily find [7] the inverse transformation from (J, Θ) to $(\phi, d\phi/dt)$,

$$\phi = 2\sin^{-1}\left[\kappa\, sn\left(2K(\kappa)\Theta/\pi, \kappa \right) \right]; \qquad \frac{d\phi}{dt} = \pm 2\kappa\sqrt{gl}\, cn\left(2K(\kappa)\Theta/\pi, \kappa \right)$$

$$(1.10)$$

where sn and cn are Jacobi elliptic functions.

For the case of rotations, there is no turning point, but one can define the action J for the running trajectory as

$$J = \frac{1}{2\pi} \int_{-\pi}^{\pi} d\phi \sqrt{2(E - gl) + gl\cos\phi} = \frac{4\sqrt{gl}}{\pi\kappa} E(\kappa_1) \qquad (1.11)$$

where the modulus $\kappa_1 = \sqrt{2gl/E}$. The angle Θ, obtained as in the previous case, is given by

$$\Theta(t) = \frac{\pi\sqrt{gl}}{\kappa_1 K(\kappa_1)} t + \Theta(0). \qquad (1.12)$$

A canonical transformation to the original variables results in

$$\phi = 2\, am\left(\frac{K(\kappa_1)\Theta}{\pi}, \kappa_1 \right); \qquad \frac{d\phi}{dt} = \pm 2\frac{\sqrt{gl}}{\kappa_1} dn\left(\frac{K(\kappa_1)\Theta}{\pi}, \kappa_1 \right)$$

$$(1.13)$$

where am is the Jacobi elliptic amplitude function, and dn is another Jacobi elliptic function.

To find the period T of the oscillating solutions, one starts from the dimensional and scaling analysis [8]. Equation (1.1) contains only one parameter, $\sqrt{l/g}$, having dimensions of time. Therefore, the ratio between T and $\sqrt{l/g}$ is dimensionless,

$$T\sqrt{\frac{g}{l}} = f(\phi). \tag{1.14}$$

For small angles, $\sin\phi \approx \phi$, and the pendulum equation (1.1) reduces to the simple equation of the harmonic oscillator with the well-known solution $T_0 = 2\pi\sqrt{l/g}$, corresponding to $f(\phi) = 2\pi$ in Eq. (1.14).

To find the function $f(\phi)$ in Eq. (1.14) for the pendulum, multiply both sides of Eq. (1.1) by $d\phi/dt$,

$$\frac{d^2\phi}{dt^2}\frac{d\phi}{dt} = \omega_0^2 \frac{d\phi}{dt}\sin\phi. \tag{1.15}$$

Integrating yields

$$\frac{1}{\omega_0}\frac{d\phi}{dt} = \sqrt{2(\cos\phi - \cos\phi_0)} \tag{1.16}$$

where ϕ_0 is the maximum value of the angle ϕ for which the angular velocity vanishes. Integrating again leads to

$$\int_0^\phi \frac{d(\phi/2)}{\left[(\sin^2(\phi/2) - \sin^2(\phi_0/2))\right]^{1/2}} = \omega_0 t \tag{1.17}$$

under the assumption that $\phi(t = 0) = 0$.

We introduce the variables ψ and k,

$$\sin\frac{\phi}{2} = k\sin\psi; \qquad k = \sin\frac{\phi_0}{2}. \tag{1.18}$$

As the angle ϕ varies from 0 to ϕ_0, the variable ψ varies from 0 to $\pi/2$. Then, (1.17) becomes an elliptic integral of the first kind $F(k, \psi)$,

$$\omega_0 t = F(k, \psi); \qquad F(k, \psi) \equiv \int_0^\psi \frac{dz}{\sqrt{1 - k^2\sin^2 z}}. \tag{1.19}$$

The rotation of the pendulum from $\phi = 0$ to $\phi = \phi_0$ takes one fourth of the period T, which is given from (1.17) by the complete elliptic integral

$F(k, \pi/2)$ of the first kind,

$$T = \frac{4}{\omega_0} F(k, \pi/2); \qquad F(k, \pi/2) \equiv \int_0^{\pi/2} \frac{dz}{\sqrt{1 - k^2 \sin^2 z}}. \qquad (1.20)$$

Since $k < 1$, one can expand the square root in (1.20) in a series and perform a term-by-term integration,

$$T = \frac{2\pi}{\omega_0} \left[1 + \left(\frac{1}{2}\right)^2 k^2 + \left(\frac{1 * 3}{2 * 4}\right)^2 k^4 + \cdots \right]. \qquad (1.21)$$

Using the power series expansion of $k = \sin(\phi_0/2)$, one can write another series for T,

$$T = \frac{2\pi}{\omega_0} (1 + \frac{\phi_0^2}{16} + \frac{11\phi_0^4}{3072} + \cdots). \qquad (1.22)$$

The period of oscillation of the plane pendulum is seen to depend on the amplitude of oscillation ϕ_0. The isochronism found by Galilei occurs only for small oscillations when one can neglect all but the first term in (1.22).

Another way to estimate the period of pendulum oscillations is by a scaling analysis [8]. In the domain $0 < t < T/4$, with characteristic angle $\phi(0) = \phi_0$, one gets the following order-of-magnitude estimates

$$\frac{d\phi}{dt} \sim -4\frac{\phi_0}{T}; \quad \frac{d^2\phi}{dt^2} \sim -16\frac{\phi_0}{T^2}; \quad \sin\phi \sim \sin\phi_0, \quad \cos\phi \sim 1. \qquad (1.23)$$

Substituting (1.23) into (1.1) yields

$$T\sqrt{\frac{g}{l}} \sim 4 \left(\frac{\phi_0}{\sin\phi_0}\right)^{1/2} \qquad (1.24)$$

or

$$\frac{T}{T_0} \sim \frac{2}{\pi} \left(\frac{\phi_0}{\sin\phi_0}\right)^{1/2} \qquad (1.25)$$

Substituting (1.23) into (1.16) rewritten as

$$\frac{1}{2} \left(\frac{d\phi}{dt}\right)^2 + \frac{g}{l} [\cos\phi_0 - \cos\phi] = 0 \qquad (1.26)$$

yields

$$\frac{T}{T_0} \sim \frac{2}{\pi} \frac{\phi_0/2}{\sin(\phi_0/2)} \qquad (1.27)$$

In the scaling analysis, one drops the numerical factor $2/\pi$ and, keeping the same functional dependencies, one writes the following general form of Eqs. (1.25) and (1.27),

$$\frac{T}{T_0} \approx \left[\frac{a\phi_0}{\sin(a\phi_0)} \right]^b. \tag{1.28}$$

Expanding $\sin(a\phi_0)$ yields

$$\frac{T}{T_0} = 1 + \frac{a^2 b}{6} \phi_0^2 + a^4 b \left(\frac{1}{180} + \frac{b}{72} \right) \phi_0^4 + \cdots . \tag{1.29}$$

From comparison with (1.22), we obtain: $a = 5\sqrt{2}/8$ and $b = 12/25$. There are also other methods of finding a and b [8].

One can easily write the solution of Eq. (1.1) for ϕ and $d\phi/dt$ in terms of elliptic integrals. Since the energy is conserved, Eq. (1.4) becomes

$$\left(\frac{d\phi}{dt} \right)^2 = -\frac{2g}{l} (1 - \cos\phi) + Const. = -\frac{4g}{l} \sin^2 \left(\frac{\phi}{2} \right) + Const. \tag{1.30}$$

Denoting the value of $(d\phi/dt)^2$ in the downward position by A, and $\sin(\phi/2)$ by y, one can rewrite (1.30) as

$$\left(\frac{dy}{dt} \right)^2 = \frac{1}{4} (1 - y^2) \left(A - \frac{4g}{l} y^2 \right). \tag{1.31}$$

Consider separately the locked and running trajectories for which the bob performs oscillations and rotations around the downward position, respectively. In the former case, $d\phi/dt$ vanishes at some $y < 1$, i.e., $Al/4g < 1$. Introducing the new positive constant k^2 by $A = 4gk^2/l$, one can rewrite Eq. (1.31) in the following form,

$$\left(\frac{dy}{dt} \right)^2 = \frac{g}{l} (1 - y^2) (k^2 - y^2). \tag{1.32}$$

The solution of this equation has the form [9]

$$y = k \, sn \left[\sqrt{g/l} \, (t - t_0), k \right] \tag{1.33}$$

where sn is the periodic Jacobi elliptic function. The two constants t_0 and k are determined from the initial conditions.

For the running solutions $Al/4g > 1$, for which $k < 1$, the differential equation (1.31) takes the following form,

$$\left(\frac{dy}{dt}\right)^2 = \frac{gl}{k^2}\left(1 - y^2\right)\left(1 - y^2 k^2\right).$$ (1.34)

The solution of this equation is

$$y = k \, sn\left[\sqrt{g/l}\,\frac{t - t_0}{k}, k\right].$$ (1.35)

Finally, for $Al = 4g$, the bob just reaches the upward position. In this case, Eq. (1.31) takes the simple form,

$$\left(\frac{dy}{dt}\right)^2 = \frac{g}{l}\left(1 - y^2\right)^2$$ (1.36)

whose solution is

$$y = \tanh\left[\sqrt{g/l}\,(t - t_0)\right].$$ (1.37)

To avoid elliptic integrals, one can use approximate methods to calculate the period T. For small ϕ, $\sin\phi \approx \phi$, and the linearized Eq. (1.1) describes the dynamics of a linear harmonic oscillator having solution $\phi = A \sin(\omega_0 t)$, with $\omega_0 = \sqrt{g/l}$. We use this solution as the first approximation to the nonlinear equation and substitute it into (1.1). To obtain a better solution, and then repeat this process again and again. In the first approximation, one obtains

$$\frac{d^2\phi}{dt^2} \approx -\omega_0^2\left[A \sin(\omega_0 t) - \frac{[A \sin(\omega_0 t)]^3}{3!} + \frac{[A \sin(\omega_0 t)]^5}{5!} + \cdots\right].$$ (1.38)

Each term in (1.38) contains harmonics that correspond to a power of $A \sin(\omega_0 t)$, i.e, the series is made up of terms that are the odd harmonics of the characteristic frequency ω_0 of the linear oscillator. The second approximation has a solution of the form $\phi = A \sin(\omega_0 t) + B \sin(3\omega_0 t)$. A complete description requires a full Fourier spectrum,

$$\phi = \sum_{l=0}^{\infty} A_{2l+1} \sin\left[(2l + 1)\,\omega_0 t\right].$$ (1.39)

Turning now to the calculation of the period T, one can use the following approximate method [10]. Since the period depends on the amplitude ϕ_0, one can write $T = T_0 f(\phi_0)$, where $T_0 = 2\pi\sqrt{g/l}$. One may rewrite Eq. (1.1) in the form,

$$\frac{d^2\phi}{dt^2} + \frac{g}{l}\psi(\phi)\phi = 0, \qquad \psi(\phi) = \left(\frac{\sin\phi}{\phi}\right) \tag{1.40}$$

and replace $\psi(\phi)$ by some $\psi(\overline{\phi})$. According to (1.18) and (1.21), $T = T_0(1 + \phi_0^2/16 + \cdots)$. Comparing the latter expression with the series expansion for $\psi(\overline{\phi})$, one sees that $\overline{\phi} = \sqrt{3}\phi_0/2$. Finally, one obtains for the first correction T_1 to the period,

$$T_1 = T_0 \left(\frac{\sin\left(\sqrt{3}\phi_0/2\right)}{\sqrt{3}\phi_0/2}\right)^{-1/2}. \tag{1.41}$$

A comparison between the approximate result (1.41) and the exact formula (1.20) shows that (1.41) is accurate to 1% for amplitudes up to 2.2 radian [10].

The addition of damping to Eq. (1.1) makes it analytically unsolvable. Assuming that the damping is proportional to the angular velocity, the equation of motion takes the form (1.2). This equation does not have an analytical solution, and we content ourselves with numerical solutions. One proceeds as follows. Equation (1.2) can be rewritten as two first-order differential equations,

$$z = \frac{d\phi}{dt}; \qquad \frac{dz}{dt} + \gamma z + \frac{g}{l}\sin\phi = 0. \tag{1.42}$$

The fixed points of these equations, where $z = dz/dt = 0$, are located at $\phi = 0$ and $\phi = \pm n\pi$. The $(d\phi/dt, \phi)$ phase plane changes from that shown in Fig. 1.2 for the undamped pendulum to that shown in Fig. 1.3 [11]. Simple linear stability analysis shows [12] that all trajectories will accumulate in the fixed points for even n, while for odd n, the fixed point becomes a saddle point, i.e., the trajectories are stable for one direction of perturbation but unstable for the other direction.

The higher derivatives $d^m\phi/dt^m$ are increasingly sensitive probes of the transient behavior and the transition from locked to running trajectories. The higher derivatives of the solutions of Eqs. (1.42) are shown [13] in the $\left(d^m\phi/dt^m, d^{m-1}\phi/dt^{m-1}\right)$ phase planes for $m \leq 5$.

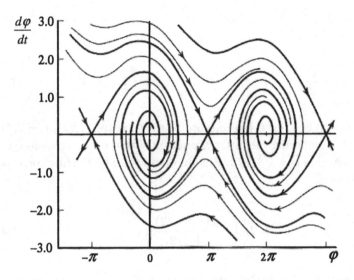

Fig. 1.3   Phase plane of a damped pendulum described by Eq. (1.2).

1.2   Isomorphic models

In some of examples given below, for simplicity we compare the derived equation with the overdamped pendulum Eq. (1.3), whereas their more complicated description will be equivalent to the underdamped Eq. (1.2).

1.2.1   *Brownian motion in a periodic potential*

By replacing the angular variable ϕ in Eq. (1.1) by the coordinate x, we obtain the equation describing one-dimensional motion of a Brownian particle in a periodic potential. The literature on this subject is quite extensive (see, for example, an entire chapter in Risken's monograph [14]).

1.2.2   *Josephson junction*

A Josephson junction consist of two weakly coupled superconductors, separated by a very thin insulating barrier. Since the size of the Cooper pair in superconductors is quite large, the pair is able to jump across the barrier producing a current, the so-called Josephson current. The basic equations governing the dynamics of the Josephson effect connects the voltage $U(t) = (\hbar/2e)(\partial\phi/\partial t)$ and the current $I(t) = I_c \sin\phi(t)$ across the Josephson junction. This defines the "phase difference" ϕ of the two superconductors across the junction. The critical current I_c is an important phenomenological parameter of the device that can be affected by temperature as well as by an applied magnetic field. The application of Kirchoff's law to this closed circuit yields,

$$I = I_c \sin\phi(t) + \frac{\hbar}{2eR}\frac{\partial\phi}{\partial t} \qquad (1.43)$$

where R is the resistivity of a circuit, and I and I_c are the bias and critical current, respectively. This equation is simply Eq. (1.3).

1.2.3 *Fluxon motion in superconductors*

The magnetic field penetrates superconductors of type-II in the form of quasi-particles called fluxons. In many cases, fluxons are moving in a periodic potential which is created by the periodic structure of pinning centers or by the plane layers of a superconductor. If one neglects the fluxon mass, the equation of fluxon motion has the form (1.3) [15].

1.2.4 *Charge density waves (CDWs)*

As a rule, at low temperatures the electron charge density is distributed uniformly in solids. The well-known violation of this rule occurs in superconducting materials where the electrons are paired. Another example of non-uniform distribution of electrons is the CDW, which behaves as a single massive particle positioned at its center of mass. CDWs have a huge dielectric constant, more than one million times larger than in ordinary materials. One can clearly see the inhomogeneity of a charge by the scanning tunneling microscope. Such a system shows "self-organization" in the sense that a small perturbation is able to induce a sudden motion of the entire charge density wave. This perturbation can be induced by an external electric field whereby an increase in voltage beyond a certain threshold value causes the entire wave to move, producing "non-Ohmic" current which vastly increases with only a small increase in voltage.

The mathematical description of the CDW based on the "single-particle" model assumes that the CDW behaves as a classical particle. The experimentally observed nonlinear conductivity and the appearance of a new periodicity form the basis for the periodically modulated lattice potential acting on the CDW, so that the equation of motion of a center of mass x of a damped CDW has the following form,

$$\frac{d^2x}{dt^2} + \gamma\frac{dx}{dt} + b\sin x = 0. \tag{1.44}$$

This equation coincides with Eq. (1.2) for a damped pendulum.

1.2.5   *Laser gyroscope*

A mechanical gyroscope with rotating wheels is widely used for orientation in space. However, nowadays they are being replaced by the laser gyroscope which works on a physical principle found by Sagnac about a hundred years ago. Sagnac found that the difference in time for two beams travelling in opposite directions around a closed path going through a rotating platform is proportional to the speed of the platform. The beam which is travelling in the direction of rotation of the platform travels for a longer distance than the counterrotating beam, and hence has a lower frequency. The phase difference ϕ between these two beams running in a ring-laser microscope, which allows one to find the velocity of a rotating platform, is satisfied by Eq. (1.3), where a denotes the rotation rate and b is the backscattering coefficient.

1.2.6   *Synchronization phenomena*

In the 17th century, the Dutch physicist Huygens found that two pendulum clocks attached to a wall, which introduces a weak coupling between them, will run at the same rate. This phenomenon of synchronization is in general present in dynamic systems with two competing frequencies. The two frequencies may arise through the coupling of an oscillator to an external periodic force. The equation which describes the influence of a small external force on the intrinsic periodic oscillations of an oscillator [16] connects the phase difference ϕ between oscillator frequency and that of an external force expressed by the frequency difference a, and the periodic force $b\sin\phi$, i.e., it has the form of Eq. (1.3) [17].

1.2.7   *Parametric resonance in anisotropic systems*

The rotation of an anisotropic cluster in an external field is described by the following equation,

$$\frac{d\mathbf{L}}{dt} = \mathbf{M} \times \mathbf{F} - \beta\boldsymbol{\omega} \tag{1.45}$$

where $\mathbf{L}$ is the angular momentum, $\boldsymbol{\omega}$ is the angular velocity of rotation, $M_i = \chi_i F_i$ is the magnetic (dielectric) moment induced in the external field $\mathbf{F}$. Anisotropy means that $\chi_1 \neq \chi_2 = \chi_3$; $\Delta\chi \equiv \chi_1 - \chi_2$, while the moment of inertia is isotropic in the $x - y$ plane, $I_1 = I_2 \equiv I$. Connecting the coordinate axis with the moving cluster, one can easily show [18] that the equation of motion for the nutation angle θ coincides with Eq. (1.2) for a damped pendulum,

$$\frac{d^2\theta}{dt^2} + \frac{\beta}{I}\frac{d\theta}{dt} + \frac{F^2\Delta\chi}{2I}\sin(2\theta) = 0. \tag{1.46}$$

For the alternating external field, $F = F_0 \cos(\omega t)$, Eq. (1.46) takes the form of the equation of motion of a pendulum with a vertically oscillating suspension point.

1.2.8 The Frenkel-Kontorova model (FK)

In its simplest form, the FK model describes the motion of a chain of interacting particles ("atoms") subject to an external on-site periodic potential [19]. This process is modulated by the one-dimensional motion of quasi-particles (kinks, breathers, etc.). The FK model was originally suggested for a nonlinear chain to describe, in the simplest way, the structure and dynamics of a crystal lattice in the vicinity of a dislocation core. Afterwards, it was also used to describe different defects, monolayer films, proton conductivity of hydrogen-bonded chains, DNA dynamics and denaturation.

1.2.9 Solitons in optical lattices

Although solitons are generally described by the sine-Gordon equation, the motion of the soliton beam in a medium with a harmonic profile of refractive index is described by the pendulum equation with the incident angle being the control parameter [20].

1.3 Noise

1.3.1 White noise and colored noise

In the following, we will consider noise $\xi(t)$ with $\langle \xi(t) \rangle = 0$ and the correlator

$$\langle \xi(t_1)\xi(t_2) \rangle = r(|t_1 - t_2|) \equiv r(z). \tag{1.47}$$

Two integrals of (1.47) characterize fluctuations: the strength of the noise D,

$$D = \frac{1}{2} \int_0^\infty \langle \xi(t) \xi(t+z) \rangle \, dz, \qquad (1.48)$$

and the correlation time τ,

$$\tau = \frac{1}{D} \int_0^\infty z \langle \xi(t) \xi(t+z) \rangle \, dz. \qquad (1.49)$$

Traditionally, one considers two different forms of noise, white noise and colored noise. For white noise, the function $r(|t_1 - t_2|)$ has the form of a delta-function,

$$\langle \xi(t_1) \xi(t_2) \rangle = 2D\delta(t - t_1), \qquad (1.50)$$

The name "white" noise derives from the fact that the Fourier transform of (1.50) is "white", that is, constant without any characteristic frequency. Equation (1.50) implies that noise $\xi(t_1)$ and noise $\xi(t_2)$ are statistically independent, no matter how near t_1 is to t_2. This extreme assumption, which leads to the non-physical infinite value of $\langle \xi^2(t) \rangle$ in (1.50), implies that the correlation time τ is not zero, as was assumed in (1.50), but smaller than all the other characteristic times in the problem.

All other non-white sources of noise are called colored noise. A widely used form of noise is Ornstein-Uhlenbeck exponentially correlated noise, which can be written in two forms,

$$\langle \xi(t) \xi(t_1) \rangle = \sigma^2 \exp\left[-\lambda |t - t_1|\right], \qquad (1.51)$$

or

$$\langle \xi(t) \xi(t_1) \rangle = \frac{D}{\tau} \exp\left[-\frac{|t - t_1|}{\tau}\right]. \qquad (1.52)$$

White noise (1.50) is characterized by its strength D, whereas Ornstein-Uhlenbeck noise is characterized by two parameters, λ and σ^2, or τ and D. The transition from Ornstein-Uhlenbeck noise to white noise (1.50) occurs in the limit $\tau \to 0$ in (1.52), or when $\sigma^2 \to \infty$ and $\lambda \to \infty$ in (1.51) in such a way that $\sigma^2/\lambda = 2D$.

A slightly generalized form of Ornstein-Uhlenbeck noise is the so-called narrow-band colored noise with a correlator of the form,

$$\langle \xi(t)\,\xi(t_1) \rangle = \sigma^2 \exp(-\lambda |t - t_1|) \cos(\Omega |t - t_1|). \qquad (1.53)$$

There are different forms of colored noise, one of which will be briefly described in the next section.

1.3.2 Dichotomous noise

A special type of colored noise with which we shall be concerned is symmetric dichotomous noise (random telegraph signal) where the random variable $\xi(t)$ may equal $\xi = \pm\sigma$ with mean waiting time $(\lambda/2)^{-1}$ in each of these two states. Like Ornstein-Uhlenbeck noise, dichotomous noise is characterized by the correlators (1.51)-(1.52).

In what follows, we will use the Shapiro-Loginov procedure [21] for splitting the higher-order correlations, which for exponentially correlated noise yields

$$\frac{d}{dt}\langle \xi \cdot g \rangle = \left\langle \xi \frac{dg}{dt} \right\rangle - \lambda \langle \xi \cdot g \rangle, \qquad (1.54)$$

where g is some function of noise, $g = g\{\xi\}$. If $dg/dt = B\xi$, then Eq. (1.54) becomes

$$\frac{d}{dt}\langle \xi \cdot g \rangle = B\langle \xi^2 \rangle - \lambda \langle \xi \cdot g \rangle, \qquad (1.55)$$

and for white noise ($\xi^2 \to \infty$ and $\lambda \to \infty$, with $\xi^2/\lambda = 2D$), one obtains

$$\langle \xi \cdot g \rangle = 2BD. \qquad (1.56)$$

1.3.3 Langevin and Fokker-Planck equations

Noise was introduced into differential equations by Einstein, Smoluchowski and Langevin when they considered the molecular-kinetic theory of Brownian motion. They assumed that the total force acting on the Brownian particle can be decomposed into a systematic force (viscous friction proportional to velocity, $f = -\gamma v$) and a fluctuation force $\xi(t)$ exerted on the Brownian particle by the molecules of the surrounding medium. The fluctuation force derives from the different number of molecular collisions with a Brownian particle of mass m from opposite sides resulting in its random motion. The motion of a Brownian particle is described by the so-called Langevin equation

$$m\frac{dv}{dt} = -\gamma v + \xi(t).$$ (1.57)

The stochastic Eq. (1.57) describes the motion of an individual Brownian particle. The random force $\xi(t)$ in this equation causes the solution $v(t)$ to be random as well. Equivalently, one can consider an ensemble of Brownian particles and ask how many particles in this ensemble have velocities in the interval $(v, v + dv)$ at time t, which defines the probability function $P(v,t)\, dv$. The deterministic equation for $P(v,t)$ is called the Fokker-Planck equation, which has the following form for white noise [22],

$$\frac{\partial P(v,t)}{\partial t} = \frac{\partial}{\partial v}(\gamma v P) + D\frac{\partial^2 P}{\partial v^2}.$$ (1.58)

In the general case, for a system whose the equation of motion $dx/dt = f(x)$ has a nonlinear function $f(x)$, the Langevin equation has the following form,

$$\frac{dx}{dt} = f(x) + \xi(t)$$ (1.59)

with the appropriate Fokker-Planck equation being

$$\frac{\partial P(x,t)}{\partial t} = -\frac{\partial}{\partial v}[f(x)P] + D\frac{\partial^2 P}{\partial v^2}.$$ (1.60)

Thus far, we have considered additive noise which describes an internal, say, thermal noise. However, there are also fluctuations of the surrounding medium (external fluctuations) which enter the equations as multiplicative noise,

$$\frac{dx}{dt} = f(x) + g(x)\xi(t).$$ (1.61)

The appropriate Fokker-Planck equation then has the form [22]

$$\frac{\partial P(x,t)}{\partial t} = -\frac{\partial}{\partial x}[f(x)P] + D\frac{\partial}{\partial x}g(x)\frac{\partial}{\partial x}g(x)P.$$ (1.62)

We set aside the Ito-Stratonovich dilemma [22] connected with Eq. (1.62).

The preceding discussion was related to first-order stochastic differential equations. Higher-order differential equations can always be written as a system of first-order equations, and the appropriate Fokker-Planck equation will have the following form

$$\frac{\partial P(x,t)}{\partial t} = -\sum_i \frac{\partial}{\partial x_i} [f_i(x) P] + \frac{1}{2} \sum_{i,j} \frac{\partial^2}{\partial x_i \partial x_j} [g_{ij}(x) P] \qquad (1.63)$$

for any functions $f_i(x)$ and $g_{ij}(x)$. The linearized version of (1.63) is

$$\frac{\partial P(x,t)}{\partial t} = -\sum_{i,j} f_{ij} \frac{\partial}{\partial x_i} (x_j P) + \frac{1}{2} \sum_{i,j} g_{ij} \frac{\partial^2 P}{\partial x_i \partial x_j} \qquad (1.64)$$

where f_{ij} and g_{ij} are now constant matrices.

For the case of colored (not-white) noise, there is no rigorous way to find the Fokker-Planck equation that corresponds to the Langevin Eqs. (1.59), (1.61), and one has to use different approximations [14].

One can illustrate [23], the importance of noise in deterministic differential equations by the simple example of the Mathieu equation supplemented by white noise $\xi(t)$

$$\frac{d^2\phi}{dt^2} + (\alpha - 2\beta \cos 2t)\phi = \xi(t). \qquad (1.65)$$

Solutions of Eq. (1.65) in the absence of noise are very sensitive to the parameters α and β, which determine regimes in which the solutions can be periodic, damped or divergent. In order to write the Fokker-Planck equation corresponding to the Langevin equation (1.65), we decompose this second-order differential equation into the two first-order equations

$$\frac{d\phi}{dt} = \Omega; \qquad \frac{d\Omega}{dt} = -(\alpha - 2\beta \cos 2t)\phi + \xi(t). \qquad (1.66)$$

The Fokker-Planck equation (1.63) for the distribution function $P(\phi, \Omega, t)$ then takes the form

$$\frac{\partial P}{\partial t} = D\frac{\partial^2 P}{\partial \Omega^2} - \Omega\frac{\partial P}{\partial \phi} + (\alpha - 2\beta \cos 2t)\phi\frac{\partial P}{\partial \Omega} \qquad (1.67)$$

with initial conditions $P(\phi, \Omega, 0) = \delta(\phi - \phi_0)\delta(\Omega - \Omega_0)$. Equation (1.67) can be easily solved by Fourier analysis to obtain the following equation for the variance $\sigma^2 \equiv \langle \phi^2 \rangle - \langle \phi \rangle^2$,

$$\frac{d^3(\sigma^2)}{dt^3} + 4(\alpha - 2\beta \cos 2t)\frac{d(\sigma^2)}{dt} + (8\beta \sin 2t)\sigma^2 = 8D. \qquad (1.68)$$

The solutions of Eq. (1.68) have the same qualitative properties as those of Eq. (1.65). However, for sufficiently large values of β, the variance does not exhibit the expected diffusional linear dependence but increases exponentially with time.

For this simple case of a linear differential equation, one can obtain an exact solution. However, the equation of motion of the pendulum is non-linear which prevents an exact solution. We will see from the approximate calculations and numerical solutions that the existence of noise modifies the equilibrium and dynamical properties of the pendulum in fundamental way.

Chapter 2

Overdamped Pendulum

2.1 Deterministic motion

The overdamped equation of motion (1.3),

$$\frac{d\phi}{dt} = a - b\sin\phi; \qquad b \equiv \frac{g}{l} \tag{2.1}$$

corresponds to motion in the washboard potential $U(\phi) = -a\phi - b\cos\phi$ shown in Fig. 2.1.

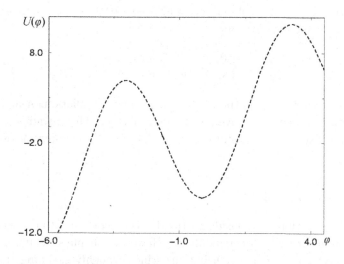

Fig. 2.1   Washboard potential $U(\phi) = a_0\phi - b_0\cos\phi$.

The two torques in Eq. (2.1) have opposite tendencies: the constant torque strives to increase ϕ (running solutions) while the second torque tends to keep ϕ at its minima (locked solutions). The solution of Eq. (2.1) depends on the comparative importance of these two factors. Equation (2.1) can be solved exactly for the simplest periodic function $a(t)$ of the square-wave pulses (periodic telegraph signal). For $a > b$, the solution of this equation has the following form [24],

$$t - t_0 = \frac{2}{\sqrt{a^2 - b^2}} \left[\tan^{-1} \left(\frac{a \tan(\phi/2) - b}{\sqrt{a^2 - b^2}} \right) \pm m \frac{\pi}{2} \right], \qquad (2.2)$$

whereas for $a < b$, the solution is

$$t - t_0 = \frac{1}{\sqrt{b^2 - a^2}} \ln \left| \frac{a \tan(\phi/2) - b \cot(\phi_c/2)}{a \tan(\phi/2) - b \tan(\phi_c/2)} \right| \qquad (2.3)$$

where m is an integer, and $\phi_c = \sin^{-1}(a/b)$.

As one can see from these equations, the solutions for $a \gtrless b$ are quite different. For $a < b$, in the limit $t \to \infty$, the solution of Eq. (2.1) asymptotically approaches the stable fixed point $\phi(t) \to \phi_s$ for all initial conditions. There is another solution of Eq. (2.1) which has the form $\phi_u = \pm 1/2 - \phi_c$, which is unstable. Since no solutions can cross the branch point, the motion is bounded. On the other hand, there is no restriction on the motion of the pendulum for $a > b$. This difference can also be seen from the expression for the average time derivative of ϕ,

$$\left\langle \frac{d\phi}{dt} \right\rangle = \begin{cases} 0, & \text{for } a < b \\ \sqrt{a^2 - b^2}, & \text{for } a > b \end{cases} \qquad (2.4)$$

which implies that for $a < b$, the pendulum performs oscillations near the downward position with zero average angular velocity, whereas for $a > b$, the pendulum executes full rotations in a clockwise or counterclockwise direction.

2.2   Influence of noise

In the previous section, we considered the deterministic Eq. (2.1) describing the dynamics of a single pendulum. However, all physical parameters are subject to random perturbations which, roughly speaking, may have internal or external origin. The former (additive noise) will influence the parameter a in Eq. (2.1), whereas the latter (multiplicative noise) are responsible for fluctuations in b,

$$a = a_0 + \xi(t); \qquad b = b_0 + \eta(t) \tag{2.5}$$

so that Eq. (2.1) takes the following form

$$\frac{d\phi}{dt} = [a_0 + \xi(t)] - [b_0 + \eta(t)] \sin\phi. \tag{2.6}$$

For example, for fluxon motion, additive noise $\xi(t)$ comes from the thermal fluctuations, whereas multiplicative noise $\eta(t)$ originates from the thermal activation of distinctive pinning centers.

The important role of noise can be illustrated as follows. The washboard potential $U(\phi) = -a\phi - b\cos\phi$, consists of wells and barriers as shown in Fig. 2.1. For small a, the motion of the particle is essentially confined to one potential well, which is equivalent to pendulum motion back and forth around the downward position. However, for large a, the particle can overcome the barrier following the driving force a (the rotation motion of a pendulum). There is a clear threshold ($a \gtrsim b$) between these "locked" and "running" states. In the presence of noise, this threshold is blurred since, even for small a, the particle is able to overcome the potential barrier to follow the driving force a. In this section we will show many other effects due to different types of noise.

2.2.1 Additive white noise

Two quantities characterize the dynamics: the average velocity

$$\left\langle \frac{d\phi}{dt} \right\rangle \equiv \lim_{t \to \infty} \frac{\phi(t)}{t} \tag{2.7}$$

and the effective diffusion coefficient

$$D_{eff} \equiv \lim_{t \to \infty} \frac{1}{2t} \left[\left\langle [\phi(t) - \langle \phi(t) \rangle]^2 \right\rangle \right]. \tag{2.8}$$

One can find the exact analytic solution for both $\langle d\phi/dt \rangle$ and D for the more general equation of the form

$$\frac{d\phi}{dt} = a - \frac{dU}{d\phi} + \xi(t) \tag{2.9}$$

where $U(x)$ is a periodic function with period L,

$$U(\phi + L) = U(\phi) \tag{2.10}$$

and $\xi(t)$ is white noise.

The general expression for the average velocity was obtained a long time ago [25]

$$\left\langle \frac{d\phi}{dt} \right\rangle = \frac{1 - \exp\left(aL/T\right)}{\int_{\phi_0}^{\phi_0+L} I_{\pm}\left(x\right)\left(dx/L\right)} \tag{2.11}$$

with

$$I_{+}\left(x\right) = D_0^{-1}\exp\left[\left(U(x) - ax\right]\int_{x-L}^{x} dy \exp\left\{-\left[(U(y) - ay\right]\right\} \tag{2.12}$$

and

$$I_{-}\left(x\right) = D_0^{-1}\exp\left\{-\left[(U(x) - ax\right]\right\}\int_{x}^{x+L} dy \exp\left\{\left[(U(y) - ay\right]\right\} \tag{2.13}$$

where D_0 is the diffusion coefficient for Eq. (2.9) without the periodic term, and $I_{\pm}\left(x\right)$ means that the index may be chosen to be either plus or minus.

The general formula for the diffusion coefficient (2.8), corresponding to Eq. (2.9), was obtained recently [26] by using the moments of the first passage time,

$$D = D_0 \frac{\int_{\phi_0}^{\phi_0+L} I_{\pm}\left(y\right) I_{+}\left(y\right) I_{-}\left(y\right)\left(dy/L\right)}{\left[\int_{\phi_0}^{\phi_0+L} I_{\pm}\left(y\right)\left(dy/L\right)\right]^3}. \tag{2.14}$$

As explained in [26], the ratio of the two integrals in (2.14) can be very large, so that the presence of the periodic potential in Eq. (2.9) can result in an increase of the diffusion coefficient, $D_0 \to D$, by 14 orders of magnitude!

This effect is not only very large but also has the opposite sign compared with the analogous effect for equilibrium processes. In the latter case, the diffusion coefficient decreases upon the addition of a periodic potential to the Brownian particle due to localization of the particle in the periodic potential [27].

Let us return to the sinusoidal form of the torque. The influence of thermal (additive) noise on the pendulum is described by the following equation,

$$\frac{d\phi}{dt} = a_0 - b_0 \sin\phi + \xi\left(t\right). \tag{2.15}$$

The exact solution of this equation for white noise of strength D is well known [25],

$$\langle \phi \rangle = \frac{\sinh\left(\pi a_0/D\right)}{\pi/D}\left|I_{\pi a_0/D}\frac{b_0}{D}\right|^{-2} \tag{2.16}$$

where $I_{\pi a_0/D}$ is the modified Bessel function of first-order with imaginary argument and imaginary index [28]. In the limit $a_0, D \ll 1$, Eq. (2.16) reduces to

$$\left\langle \frac{d\phi}{dt} \right\rangle = 2\sinh\frac{\pi a_0}{D}\exp\left(-\frac{2b_0}{D}\right). \tag{2.17}$$

Each of the two factors in Eq. (2.17) has a clear physical meaning [29]. The Arrenius exponential rate, $\exp\left[-\left(2b_0/D\right)\right]$, increases with D, which makes it easier for the system to overcome a potential barrier, while the pre-exponential factor, the difference between approach to the left well and to the right well, decreases as D increases, which makes the system more homogeneous.

Apart from the sinusoidal form of periodic function $U\left(\phi\right)$ in Eqs. (2.11) and (2.14), an analysis was performed [30], for the sawtooth potential. As was the case for the periodic potential, the diffusion coefficient D increases as a function of the tilting force a. Hence, there are two sources for an increase in diffusion, the effect of the tilting force a ("passive channel" in terminology of [30]) and a huge enhancement coming from the periodic force ("active channel"). An additional quantity that was studied [30] is the factor of randomness $Q = 2D/\left(L\langle d\phi/dt\rangle\right)$, which defines the relation between the diffusive and directed components in the Brownian motion.

2.2.2 *Additive and multiplicative white noise*

The Stratonovich interpretation of the Fokker-Planck equation for the probability distribution function $P\left(\phi,t\right)$, corresponding to the Langevin Eq. (2.6) with white noises $\xi\left(t\right)$ and $\eta\left(t\right)$ of strength D_1 and D_2, respectively, has the following form [31],

$$\frac{\partial P}{\partial t} = -\frac{\partial}{\partial\phi}\left[a_0 - \left(b_0 + D_2\cos\phi\right)\sin\phi\right]P + \frac{\partial^2}{\partial\phi^2}\left[\left(D_1 + D_2\sin^2\phi\right)\right]P \equiv -\frac{\partial J}{\partial\phi} \tag{2.18}$$

where J is the flux proportional to $\langle d\phi/dt\rangle$, namely, $\langle d\phi/dt\rangle = 2\pi J$.

For the stationary case, $\partial P/\partial t = 0$, one finds the following differential equation for the stationary distribution function [32]

$$\frac{dP_{st}}{dx} + \Gamma(\phi) P_{st} = J\Omega^2(\phi), \qquad (2.19)$$

where

$$\Gamma(\phi) = \frac{a_0 - b_0 \sin\phi - D_2 \sin\phi \cos\phi}{(D_1 + D_2 \sin^2\phi)}; \qquad \Omega(\phi) = (D_1 + D_2 \sin^2\phi)^{-\frac{1}{2}} \qquad (2.20)$$

The solution of the first-order differential equation (2.19) contains one constant which, together with the second constant J, are determined from the normalization condition, $\int_{-\pi}^{\pi} P(\phi) d\phi = 1$, and the periodicity condition, $P(-\pi) = P(\pi)$. The exact expression for P_{st} can be accurately represented by its approximate form obtained by the method of steepest descent [33],

$$P_{st} \simeq C \left[\left(1 + \frac{D_2}{D_1} \sin^2\phi \right)^{1/2} + (a_0 + b_0 \sin\phi)/D_1 \right]^{-1} \qquad (2.21)$$

where C is the normalization constant. As can be seen from Eq. (2.21), for $D_2 < D_1$, the term $a_0 + b_0 \sin\phi$ makes the main contribution to P_{st}. In this case, for $|a_0| = |b_0|$, P_{st} contains a single maximum. For $D_2 > D_1$, the main contribution to P_{st} is the term $\left[1 + (D_2/D_1) \sin^2\phi \right]^{1/2}$, i.e. P_{st} has maxima at the points $n\pi$ for integer n. Additive and multiplicative noise have opposite influence on P_{st} [33]. An increase of multiplicative noise leads to the increase and narrowing of the peaks of P_{st}, whereas the increase of additive noise leads to their decrease and broadening.

Calculations yields [34]

$$\left\langle \frac{d\phi}{dt} \right\rangle = \frac{2\pi \left\{ 1 - \exp\left[-2\pi a_0/\sqrt{D_1(D_1 + D_2)} \right] \right\}}{\left[\int_0^{2\pi} \Omega(x) F(x,0) \left(\int_x^{x+2\pi} \Omega(y) F(0,y) dy \right) dx \right]} \qquad (2.22)$$

where

$$F(k,l) = \exp\left[-\int_k^l T(z) dz \right]; \qquad T(z) = \frac{a_0 - b_0 \sin z}{D_1 + D_2 \sin^2 z}. \qquad (2.23)$$

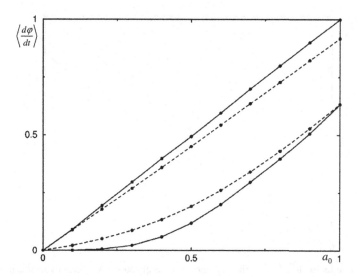

Fig. 2.2 Average angular velocity as a function of bias for $b_0 = 1$. Solid and dotted lines describe single multiplicative noise and single additive noise, respectively. The upper and lower curves correspond to noise of strength 2 and 0.1, respectively.

We have performed [34] numerical calculations of $\langle d\phi/dt \rangle$ in order to compare the importance of additive and multiplicative noise. The average angular velocity $\langle d\phi/dt \rangle$ and a_0 correspond to the voltage and the bias current for a Josephson junction. Therefore, the graph $\langle d\phi/dt \rangle$ versus a_0 gives the voltage-current characteristic of a junction. In Fig. 2.2, we present $\langle d\phi/dt \rangle$ as a function of a_0 for $b_0 = 1$, in the presence of one of the two noises given by $D = 0.1$ and $D = 2.0$. From this figure one can see that for a small value of noise ($D = 0.1$), additive noise leads to higher flux than for multiplicative noise, whereas for a larger value of noise ($D = 2$), the opposite result occurs. The transition takes place for an intermediate value of noise. As shown in Fig. 2.3, for noise of strength $D = 1$ (of order b_0), additive noise produces a larger average angular velocity for small a_0 and smaller flux for larger a_0.

For the following analysis, it is convenient to consider separately the two limiting cases of weak ($D_1 \to 0$) and strong ($D_1 \to \infty$) additive noise, combining analytic and numerical calculations. Let us start with the case of weak noise, $D_1 \to 0$ and $D_2 \to 0$, where both $D_2 > D_1$ and $D_2 < D_1$ are possible. One can use the method of steepest descent to calculate the

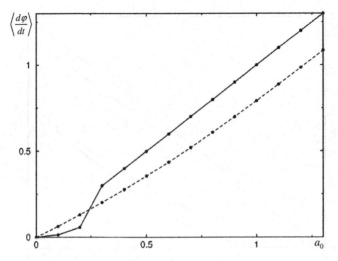

Fig. 2.3   Same as for Fig. 2.2 in the presence of a single source of noise of strength 1.

integrals in Eq. (2.22),

$$\left\langle \frac{d\phi}{dt} \right\rangle = \left[1 - \exp\left(\frac{-2\pi a_0}{\sqrt{D_1 \left(D_1 + D_2 \right)}} \right) \right] \frac{\sqrt{\left| T\left(z_{\max} \right) \dot{T}\left(z_{\min} \right) \right|}}{\Omega \left(z_{\max} \right) \Omega \left(z_{\min} \right)}$$

$$\times \exp\left[\int_{z_{\max}}^{z_{\min}} T\left(z \right) dz \right] \tag{2.24}$$

where $z_{\min}$ and $z_{\max}$ are two neighboring zeros of $T(z)$ with $T(z_{\max}) > 0$, $T(z_{\min}) < 0$.

It is easily found from Eq. (2.23) that $\sin(z_{\max,\min}) = a_0/b_0$, $\cos z_{\max} = \omega/b_0$, $\cos z_{\min} = -\omega/b_0$, and, for $b_0 > a_0$, Eq. (2.24) reduces to

$$\left\langle \frac{d\phi}{dt} \right\rangle = \sqrt{b_0^2 - a_0^2} \left[1 - \exp\left(-\frac{2\pi a_0}{\sqrt{D_1 \left(D_1 + D_2 \right)}} \right) \right] \exp \int_{z_{\max}}^{z_{\min}} T\left(z \right) dz. \tag{2.25}$$

One can evaluate the integral in (2.25). However, instead of writing this cumbersome expression, we present the results for the two limiting cases of large and small multiplicative noise compared with additive noise, $D_2 \lessgtr D_1$.

For $D_2 < D_1$, i.e., for weak additive and no multiplicative noise, one obtains the well-known result [35],

$$\left\langle \frac{d\phi}{dt} \right\rangle_{D_2 < D_1} = \sqrt{b_0^2 - a_0^2} \exp\left(\frac{\pi a_0}{D_1}\right) \exp\left[-\frac{2\sqrt{b_0^2 - a_0^2}}{D_1} - \frac{2a_0}{D_1} \sin^{-1} \frac{a_0}{b_0}\right]$$

(2.26)

whereas for weak noise with $D_2 > D_1$,

$$\left\langle \frac{d\phi}{dt} \right\rangle_{D_2 > D_1} = \sqrt{b_0^2 - a_0^2} \exp\left(\frac{-\pi a_0}{\sqrt{D_1 D_2}}\right) \left(\frac{b_0 - \sqrt{b_0^2 - a_0^2}}{b_0 + \sqrt{b_0^2 - a_0^2}}\right)^{\frac{b_0}{D_2}}.$$

(2.27)

Comparing Eqs. (2.26) and (2.27) shows that adding multiplicative noise leads to an increase in the average angular velocity in a system subject only to weak additive noise. These analytic results are supported by a numerical analysis of Eq. (2.22), given in Fig. 2.4 for small $D_1 = 0.1$ and for different values of D_2, which shows the strong influence of multiplicative noise on the flux for a small driving force.

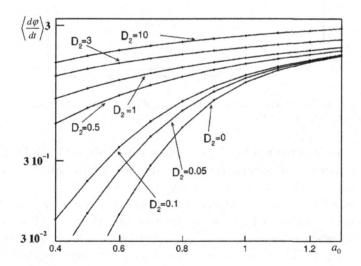

Fig. 2.4 Average angular velocity as a function of bias for $b_0 = 1$ and $D_1 = 0.1$ for different values of D_2.

Turning now to the opposite limiting case of strong additive noise, $D_1 \to \infty$, one can substantially simplify Eq. (2.22) to the following form,

$$\left\langle \frac{d\phi}{dt} \right\rangle_{D_1 \to \infty} = \frac{a_0 \pi^2}{D_1} \left(1 + \frac{D_2}{D_1} \right)^{-1/2} \left[\int_0^\pi \Omega(z) \, dz \right]^{-2}. \qquad (2.28)$$

Figure 2.5 shows the dimensionless average angular velocity $(1/a_0) \langle d\phi/dt \rangle_{D_1 \to \infty}$ as a function of D_2/D_1 for large additive noise D_1. The curve starts from $(1/a_0) \langle d\phi/dt \rangle_{D_1 \to \infty} = 1$ for $D_2 = 0$ (large additive noise suppresses the sin term in Eq. (2.6), yielding Ohm's law for the Josephson junction [35]), and increases markedly as the strength of multiplicative noise increases.

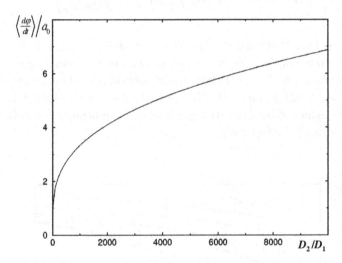

Fig. 2.5   Dimensionless average angular velocity $\langle d\phi/dt \rangle$ as a function of the ratio of noise strength D_2/D_1.

Figure 2.6 shows the results of the numerical analysis of Eq. (2.22) for comparable values of all parameters (b_0, D_1 and D_2), which again demonstrates an increase of the flux due to multiplicative noise.

One concludes that in the presence of one source of noise, the average angular velocity $\langle d\phi/dt \rangle$ is larger for additive noise if the strength of the noise is small, whereas for strong noise, multiplicative noise is more effective (Figs. 2.4 and 2.6). The transition regime between these two cases occurs for noise strength of order b_0 (the critical current for a Josephson junction), where additive noise is more effective for small driving forces and less

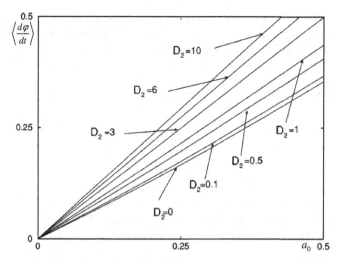

Fig. 2.6 Dimensionless average angular velocity $\langle d\phi/dt \rangle$ as a function of bias for $b_0 = 1, D_1 = 1$ for different values of D_2.

effective than multiplicative noise for large driving forces (Fig. 2.2). In fact, for small noise strength (say, $D = 0.1$) and small a_0, multiplicative noise produces flux larger by many orders of magnitude than the flux caused by additive noise. It is not surprising that multiplicative noise becomes important when D is of the order of the potential barrier height b_0.

If both sources of noise are present, then the flux is essentially increased in the presence of strong multiplicative noise for weak (Fig. 2.4), strong (Fig. 2.5) and intermediate (Fig. 2.6) strength of additive noise, especially for small bias force a_0. The latter result has a simple intuitive explanation. Indeed, the horizontal periodic potential ($a_0 = 0$) with strong fluctuations in the width of this potential has no preferred direction and, therefore, the average angular velocity vanishes, $\langle d\phi/dt \rangle = 0$. It is sufficient to have a small slope of a periodic potential, $a_0 \neq 0$, for $\langle d\phi/dt \rangle \neq 0$ to occur.

The importance of multiplicative noise for stationary states has long been known [36]. The influence of both additive and multiplicative noise on the escape time from a double-well potential was studied in [37] and [38]. The analysis of the stationary probability distribution function for a periodic potential and dichotomous multiplicative noise was given by Park et al. [39]. We have studied [32] the influence of both additive and multiplicative noise on the voltage-current characteristics of Josephson junctions. A similar effect for the output-input relation for motion in a double-well potential has been studied intensively by two groups of researchers, who called this effect "noise-induced hypersensitivity" [40] and "amplification of weak signals via on-off intermittency" [41].

2.2.3   Additive dichotomous noise

For the pendulum subject to a pure periodic torque ($a = 0$ in Eq. (2.15)), one obtains

$$\frac{d\phi}{dt} + b_0 \sin \phi = \xi(t) \tag{2.29}$$

where $\xi(t)$ is asymmetric dichotomous noise described in Sec. 1.3.2 ($\xi = A$ or $\xi = -B$, with transition probabilities λ_1 and λ_2). The asymmetry can be described by the parameter ε [42], so that

$$A = \sqrt{\frac{D}{\tau}\left(\frac{1+\varepsilon}{1-\varepsilon}\right)}; \qquad B = \sqrt{\frac{D}{\tau}\left(\frac{1-\varepsilon}{1+\varepsilon}\right)}; \qquad \gamma_{1,2} = \frac{1 \pm \varepsilon}{2}. \tag{2.30}$$

Equations (2.30) satisfy the condition $\gamma_1 B = \gamma_2 A$. Therefore, the requirement $<\xi(t)> = 0$ is obeyed.

In the presence of dichotomous noise, it is convenient to define two probability densities, $P_+(\phi, t)$ and $P_-(\phi, t)$, which correspond to the evolution of $\phi(t)$ subject to noise of strength A and $-B$, respectively. The set of Fokker-Planck equations satisfied by these two functions is a slight generalization of that considered in Sec. 1.3.3,

$$\frac{\partial P_+}{\partial t} = -\frac{\partial}{\partial \phi}\left[(b_0 \sin \phi + A) P_+\right] - \gamma_1 P_+ + \gamma_2 P_-, \tag{2.31}$$

$$\frac{\partial P_-}{\partial t} = -\frac{\partial}{\partial \phi}\left[(b_0 \sin \phi - B) P_-\right] - \gamma_2 P_- + \gamma_1 P_+. \tag{2.32}$$

Introducing the probability function $P = P_+ + P_-$, which satisfies the normalization and periodicity conditions, one obtains the cumbersome expression for the average angular velocity [42] which, to first order in the parameter $\sqrt{\tau/D}$, takes the following form

$$\left\langle \frac{d\phi}{dt} \right\rangle = \frac{\varepsilon}{\sqrt{1-\varepsilon^2}} \frac{\sqrt{\tau/D}}{I_0^2(b_0/D)} \tag{2.33}$$

where I_0 is a modified Bessel function.

Another limiting case [32] where the solution has a simple form is for slow jumps, $\gamma_{1,2} \to 0$ ("adiabatic approximation"),

$$\left\langle \frac{d\phi}{dt} \right\rangle \approx \begin{cases} 0, & \text{for } a_0 - B < b, \ a_0 + A < b \\ \frac{1}{\gamma_1 + \gamma_2}\left[\gamma_1 \sqrt{(a_0+A)^2 - b^2} + \gamma_2 \sqrt{(a_0-B)^2 - b^2}\right], & \text{for } a_0 - B > b, \ a_0 + A > b \\ \frac{\gamma_2}{\gamma_1 + \gamma_2}\sqrt{(a_0+A)^2 - b^2}, & \text{for } a_0 - B < b, \ a_0 + A > b. \end{cases}$$

$$(2.34)$$

These equations imply that in the adiabatic approximation, the total mobility is the average of the mobilities for the two corresponding potentials [43].

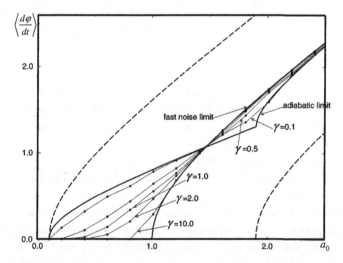

Fig. 2.7 Dimensionless average angular velocity $\langle d\phi/dt \rangle$ as a function of bias with additive dichotomous noise for different values of noise rate $\gamma = \gamma_1 + \gamma_2$. Parameters are $b_0 = 1$ and $A = B = 0.9$.

The typical average flux-bias force curves for a system subject to additive dichotomous noise are shown by the dotted line in Fig. 2.7, and — for different values of the parameters — in Figs. 2.8 and 2.9. As expected, the presence of noise smears out the sharp threshold behavior defined in (2.4), with the smearing depending on both the noise amplitude and rate. In Fig. 2.9, we show the average flux $\langle d\phi/dt \rangle$ as a function of noise rate

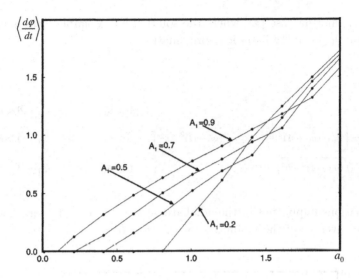

Fig. 2.8   Same as for Fig. 2.7, but for different values of noise amplitude A. Parameters are $b_0 = 1$ and $\gamma = 0.1$.

$\gamma = \gamma_1 + \gamma_2$. This non-monotonic behavior is a special manifestation of stochastic resonance which was found for a bistable potential by Doering and Gadoua [44]. Note that in our case, this phenomenon occurs in a very narrow region of values of the bias force a.

Several conclusions can be drawn from the graphs:

1) The average flux $\langle d\phi/dt \rangle$ does not vanish even for zero bias force. This phenomenon is a special case of the more general "ratchet effect" for which the net transport is induced by nonequilibrium fluctuations when some asymmetry is present. These general conditions are satisfied in our case of asymmetric dichotomous noise. One can easily verify that the ratchet effect disappears in the limiting case of symmetric noise when $a = 0$, which leads to the vanishing of $\langle d\phi/dt \rangle$. The latter case occurs both for white noise and for symmetric dichotomous noise, since the function $\Gamma(\phi)$ defined in (2.20) is an odd function for $a = 0$, which implies that $\langle d\phi/dt \rangle = 0$. The ratchet effect might have practical applications for superconducting electronics, as well as in other fields of physics, chemistry and biology (some recent references can be found in [45]).

2) A stochastic resonance phenomenon (non-monotonic behavior of the average flux as a function of noise rate) has been found in a narrow regime of the bias force, as shown in Fig. 2.9.

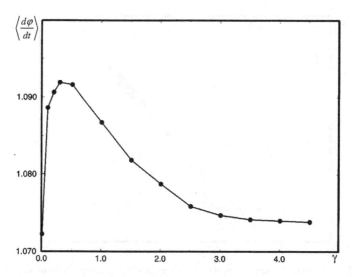

Fig. 2.9 Dimensionless average angular velocity $\langle d\phi/dt \rangle$ as a function of noise rate γ. Parameters are $b_0 = 1$ and $A = B = 0.9$.

2.2.4 *Multiplicative dichotomous noise*

Analogous to (2.34), the average angular velocity takes a simple form in the limiting case of slow jumps, γ_1, $\gamma_2 \to 0$ [32],

$$\left\langle \frac{d\phi}{dt} \right\rangle \sim \begin{cases} \frac{\gamma_2 \sqrt{a_0^2-(b_0+A)^2}}{\gamma_1+\gamma_2} + \frac{\gamma_1 \sqrt{a_0^2-(b_0-B)^2}}{\gamma_1+\gamma_2}, & \text{for } a_0 > b_0 + A, \ a_0 > b_0 - B \\ \frac{\gamma_1 \sqrt{a_0^2-(b_0-B)^2}}{\gamma_1+\gamma_2}, & \text{for } a_0 < b_0 + A, \ a_0 > b_0 - B \\ 0, & \text{for } a_0 < b_0 + A, \ a_0 < b_0 - B. \end{cases}$$

$$(2.35)$$

Typical graphs of flux-bias characteristics for multiplicative dichotomous noise are shown in Figs. 2.10, 2.11 and 2.12. The first two graphs show $\langle d\phi/dt \rangle$ as a function of a_0, which is characteristic of multiplicative dichotomous noise for different amplitudes and different noise rates, respectively. The non-monotonic behavior of $\langle d\phi/dt \rangle$ as a function of noise rate γ is shown in Fig. 2.12. Just as for additive noise, stochastic resonance occurs in a narrow regime of the bias a_0.

The non-trivial type of ratchets which exists in this case show up at the onset of the average angular velocity in the absence of a bias force as a result of the nonequilibrium noise [46]. In the absence of additive noise with symmetric multiplicative noise, $f_2(t) = \pm b$, our basic equation has

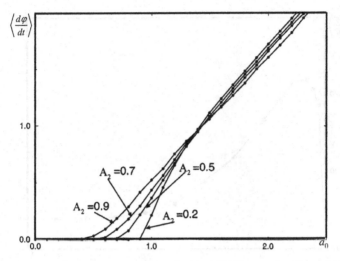

Fig. 2.10   Dimensionless average angular velocity $\langle d\phi/dt \rangle$ as a function of bias with multiplicative dichotomous noise for different noise amplitudes.

the following form,

$$\frac{d\phi}{dt} = a_0 - f_2(t)\sin\phi. \tag{2.36}$$

If $a_0 < |b|$, Eq. (2.36) gives $\langle d\phi/dt \rangle = 0$ for both $f_2(t) = \pm b$. However, if one allows switching between two dynamics laws, the resulting motion will have a net average angular velocity. This can be seen from Fig. 2.13, where the two washboard potentials $V_\pm = a\phi \pm b\cos\phi$ are shown. If, as usually assumed, the rate of reaching the minimal energy in each well is much larger than γ (adiabatic approximation), $\langle d\phi/dt \rangle$ is non-zero for the following reason: a particle locked in the potential minimum 1, switches to point 2, then rapidly slides down to point 3, switches to 4, slides to 5, etc.

2.2.5   *Joint action of multiplicative noise and additive noise*

In addition to the analysis of two sources of white noise performed in Sec. 2.2.2, there is a special case in which one or both sources of noise are dichotomous. We consider in more detail this special case in which the multiplicative noise $\eta(t)$ in Eq. (2.6) is dichotomous and the noise $\xi(t)$ is additive [47], [39],

$$\langle \eta(t)\eta(t+\tau) \rangle = \Delta^2 \exp(-2\lambda|\tau|); \quad \langle \xi(t)\xi(t+\tau) \rangle = 2\sigma^2\delta(\tau). \tag{2.37}$$

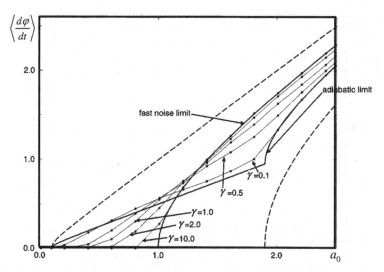

Fig. 2.11 Same as for Fig. 2.10, but for different values of noise rates γ. Parameters are $b_0 = 1$ and $A = B = 0.9$.

In the presence of dichotomous noise, it is convenient to define two probability densities, $P_+(\phi, t)$ and $P_-(\phi, t)$ which correspond to the evolution of $\phi(t)$ subject to noise of strength Δ and $-\Delta$, respectively. As it was already used in (2.31)–(2.32), the set of Fokker-Planck equations satisfied by these two functions has the following form,

$$
\begin{aligned}
\frac{\partial P_+}{\partial t} &= -\frac{\partial}{\partial \phi} \left\{ [a_0 - (b_0 + \Delta) \sin \phi] P_+ - \sigma^2 \frac{\partial P_+}{\partial \phi} \right\} - \lambda (P_+ - P_-), \\
\frac{\partial P_-}{\partial t} &= -\frac{\partial}{\partial \phi} \left\{ [a_0 - (b_0 - \Delta) \sin \phi] P_- - \sigma^2 \frac{\partial P_-}{\partial \phi} \right\} + \lambda (P_+ - P_-).
\end{aligned}
\tag{2.38}
$$

Equations (2.38) can be replaced by the set of equations for $P = P_+ + P_-$ and $Q = P_+ - P_-$,

$$
\begin{aligned}
\frac{\partial P}{\partial t} &= \frac{\partial}{\partial \phi} \left\{ [-a_0 + b_0 \sin \phi] P - \Delta Q \sin \phi - \sigma^2 \frac{\partial P}{\partial \phi} \right\} \equiv -\frac{\partial J}{\partial \phi}, \\
\frac{\partial Q}{\partial t} &= \frac{\partial}{\partial \phi} \left\{ [-a_0 + b_0 \sin \phi] Q - \Delta P \sin \phi - \sigma^2 \frac{\partial Q}{\partial \phi} \right\} - 2\lambda Q.
\end{aligned}
\tag{2.39}
$$

In the limit of $\Delta, \lambda \to \infty$ with $\Delta^2/2\lambda$ fixed and $\Delta^2/2\lambda \equiv D$, we recover the results considered previously for two sources of white noise. For $a_0 = 0$,

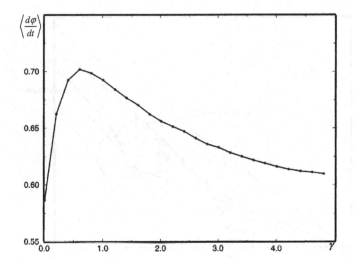

Fig. 2.12   Dimensionless average angular velocity $\langle d\phi/dt \rangle$ as a function of noise rate γ for multiplicative dichotomous noise. Parameters are $b_0 = 1$ and $A = B = 0.9$.

the stationary probability density $P_{st}(\phi)$ has the following form

$$P_{st}(\phi) = C \frac{1}{\sqrt{1 + \kappa^2 \sin \phi}} \left[\frac{\sqrt{1 + \kappa} + \sqrt{\kappa} \cos \phi}{\sqrt{1 + \kappa} - \sqrt{\kappa} \cos \phi} \right]^{b_0/2D\sigma\sqrt{1+\kappa}} \tag{2.40}$$

where $\kappa = D^2/\sigma^2$ and C is the normalization factor. When the strength of the multiplicative noise is small, $D < D_{cr}$, $P_{st}(\phi)$ has a simple maximum at $\phi = 0$. For $D > D_{cr}$, $P_{st}(\phi)$ has double maxima at $\phi = 0, \pi$ and a minimum at $\phi = \cos^{-1}(-b_0/D)$. The qualitative behavior of $P_{st}(\phi)$ does not depend on the strength σ^2 of the additive noise.

For both multiplicative dichotomous noise and additive white noise, one has to solve Eqs. (2.38) numerically. Such calculations have been performed [39] for $\sigma = 0.1$. For given b_0 and Δ, $P_{st}(\phi)$ undergoes a phase transition from the double-peak state to the single-peak state upon increasing the correlation time λ. The physical explanation of such behavior is as follows: for fast processes, the system is under the influence of average noise and, therefore, it tends to be attracted to the fixed point $\phi = 0$ (single-peak state). For slow processes and strong noise strength, the system spend most of its time at the two fixed points $\phi = 0$ and $\phi = \pi$ (double-peak state).

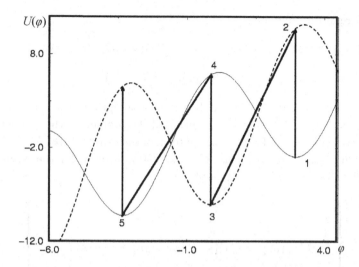

Fig. 2.13 Washboard potential $U(\phi) = a\phi - b\cos\phi$, with $a = 1$ and $b = 7$. A particle cannot move along each of the potentials by itself, but if one allows a switching between potentials, the particle moves downhill along the trajectory $1:2:3:4:5$.

A more detailed analysis has been performed [32]. We bring only the main results:

1. Two sources of dichotomous noise are able to produce a flux for small bias, whereas each by itself is unable to produce non-zero flux in this region of bias. This effect occurs also for two sources of white noise.

2. The simultaneous action of two sources of noise can be larger than each source by itself in some region of the average flux-bias plane, but smaller in other regions.

3. The "ratchet effect" occurs for asymmetric additive noise in the presence of multiplicative noise. The latter eases the requirements for the onset of the ratchet.

2.2.6 *Correlated additive noise and multiplicative noise*

Thus far, we have considered additive noise and multiplicative noise as being independent. Correlations between different sources of noise may occur when they both have the same origin, as in laser dynamics [48], or when strong external noise leads to an appreciable change in the internal structure of the system and hence its internal noise. For the simplest case of the two sources of white noise, $\xi(t)$ and $\eta(t)$ in Eq. (2.6) with the same type of correlations, one obtains

$$\langle \xi (t_1) \xi (t_2) \rangle = 2D_1 \delta (t_1 - t_2) ; \quad \langle \eta (t_1) \eta (t_2) \rangle = 2D_2 \delta (t_1 - t_2) ;$$
$$\langle \xi (t_1) \eta (t_2) \rangle = 2\lambda \sqrt{D_1 D_2} \delta (t_1 - t_2) \tag{2.41}$$

where the coefficient $\lambda (0 \leq \lambda \leq 1)$ indicates the strength of the correlation. A simple calculation [49] yields the solution of the problem with correlated multiplicative noise and additive noise in term of the results (2.18)-(2.23) for a system with non-correlated noise (see also [50]). Finally, for the steady-state average angular velocity, one obtains

$$\left\langle \frac{d\phi}{dt} \right\rangle = \left\{ \int\limits_{0}^{2\pi} \frac{d\phi_1 \int\limits_{\phi_1}^{\phi_1 + 2\pi} \exp\left[\Psi (\phi_2) \right] d\phi_2}{B (\phi_1) \exp\left[\Psi (\phi_1) \right] - B (\phi_1 + 2\pi) \exp\left[\Psi (\phi_1 + 2\pi) \right]} \right\}^{-1} , \tag{2.42}$$

where

$$\Psi (z) = - \int\limits_{0}^{z} dx \, A (x) / B (x) ;$$

$$B (x) = D_2 \sin^2 x - 2\lambda \sqrt{D_1 D_2} \sin x + D_1 ; \tag{2.43}$$

$$A (x) = a_0 - b_0 \sin x - \lambda \sqrt{D_1 D_2} \cos x + D_2 \sin x \cos x.$$

Extensive numerical calculations have been carried out [50] for non-correlated noise ($\lambda = 0$) which lead to the conclusions described in Sec. 2.2.5, and for $-1 \leq \lambda < 0$ and $0 < \lambda \leq 1$, which lead to new phenomena:

1. **Reversal of** $\langle d\phi/dt \rangle$. For non-zero values of λ, the direction of $\langle d\phi/dt \rangle$ reverses when the ratio D_1/D_2 decreases.

2. **Existence of extremum.** As D_1/D_2 increases, $\langle d\phi/dt \rangle$ possesses a minimum changing from negative to positive values for $\lambda > 0$, and a maximum changing from positive to negative values for $\lambda < 0$. Both maximum and minimum exist for completely correlated noise, $\lambda = 1$.

3. **Symmetric dependence of** $\langle d\phi/dt \rangle$ **as a function of** $\mathbf{D_1/D_2}$. Calculations of $\langle d\phi/dt \rangle$ have also been performed [51] for the case where the multiplicative white noise $\eta (t)$ in (2.41) was replaced by dichotomous noise.

The following general comment should be made. The non-zero steady-state angular velocity is obtained from Eq. (2.29) which does not contain any driving force. Such noise-induced motion cannot appear in an equilibrium system, which includes symmetric thermal noise, due to the second law of thermodynamics. A non-zero average velocity exists only in the case of asymmetric thermal noise, as we have seen in Sec. 2.2.3. Moreover, even in a non-equilibrium state, a non-zero average velocity can appear only in the presence of some symmetry breaking, for the following reason. If $\phi(t)$ is the solution of the dynamic equation for a given realization of noise, then $-\phi(t)$ is also a solution for t replaced by $-t$. These two solutions will give two equal average velocities equal to $\pm \langle d\phi/dt \rangle$, which implies $\langle d\phi/dt \rangle = 0$. A non-zero value can be obtained when either the potential energy or the noise is asymmetric. As we have seen, another possibility is correlation between additive and multiplicative noise.

2.3 Periodic driven force

2.3.1 *Deterministic equation*

Equation (1.3) with $a = 0$, supplemented by a external periodic force $f \sin(\omega t)$, has the following form,

$$\frac{d\phi}{dt} + b\sin\phi = f\sin(\omega t). \tag{2.44}$$

This equation has no analytic solution, but the numerical solution shows [52] that there are two types of stable periodic motion, the normal and inverted modes of oscillations which are symmetric with $\langle \phi \rangle = 0$ and $\langle \phi \rangle = \pi$, respectively.

Assuming that the solution of Eq. (2.44) is given by the trial function

$$\phi = A_0 + A_1 \cos(\omega t + \alpha_1), \tag{2.45}$$

a small perturbation $\delta(t)$ to (2.45) is governed by the linearized equation

$$\frac{d\delta}{dt} + b\delta\cos\phi = 0 \tag{2.46}$$

with the solution

$$\delta = \exp\left(-b \int dt \, \cos\phi\right). \tag{2.47}$$

Substituting (2.45) into (2.47), and using the well-known expansion of trigonometric functions in series in Bessel functions $J_n(z)$, one obtains a stability condition,

$$bJ_0\left(A_1\right)\cos A_0 > 0. \tag{2.48}$$

The boundaries between two types of solutions of Eq. (2.44), which follow from Eq. (2.48), are in good agreement with the numerical calculations [52].

Adding a constant term in the right hand side of Eq.(2.45) yields

$$\frac{d\phi}{dt} + b\sin\phi = a + f\sin\left(\omega t\right), \tag{2.49}$$

has been studied both numerically [52] and by means of perturbation theory [53]. A most remarkable result of both calculations is the lock-in phenomenon which, in the language of Josephson junctions, manifests itself as horizontal "Shapiro steps" in the voltage-current characteristics. This phenomenon is due to the stability of periodic orbits in the $(\phi, d\phi/dt)$ plane with respect to small changes of the parameter a in Eq. (2.49). Therefore, one needs large changes in a to destroy this periodic orbit and send the system to the next stable orbit, thereby producing successive Shapiro steps. In fact, Eq. (2.4) defines the zeroth Shapiro step.

One can say [54] that the Shapiro steps are a special case of synchronization with the resonance condition on two frequencies in the problem, $\langle d\phi/dt\rangle = n\omega$ for integer n, when a phase tends to synchronize its motion with the period of an external field to overcome an integer number of wells during one cycle of the force.

2.3.2 *Influence of noise*

In the presence of noise, Eq. (2.49) takes the form

$$\frac{d\phi}{dt} + b\sin\phi = a + f\sin\left(\omega t\right) + \xi\left(t\right). \tag{2.50}$$

The numerical solution of Eq. (2.50) shows [54] that the dependence of the average velocity defined in (2.11) on the strength of noise depends crucially on the value of the constant torque a. For different values of a, this dependence can increase monotonically or decrease monotonically or become non-monotonic. The presence of noise leads to the blurring of the Shapiro steps, helping a system to leave the locked-in states, and to the appearance of multiple peaks in the dependence of the effective diffusion on the constant torque a [54].

Equation (2.50) describes, along with an overdamped pendulum, the motion of a Brownian particle in a periodic potential. A fascinating analysis of the diffusion of a particle, subject to a square wave periodic force, has been performed [55]. The diffusion coefficient for Eq. (2.50) was compared with that of a free Brownian particle ($b = f = 0$). Both the analytic calculation for a square-wave periodic force and the numerical simulations for the sinusoidal potential show that the diffusion rate can be greatly enhanced compared with free motion by the appropriate choice of coordinate and time periodicity. In addition, the diffusion exhibits a maximum for a specific value of the strength of noise, thereby showing a stochastic resonance-type effect.

2.3.3 *Deterministic telegraph signal*

Since Eqs. (2.44) and (2.49) with a sinusoidal force, $f \sin(\omega t)$, do not allow analytic solutions, we consider a simplified form with a periodic telegraph signal which does allow an exact solution. This simplified equation has the following dimensionless form,

$$\frac{d\phi}{d\tau} + \sin\phi = a + a_1\psi\left(\tau/b\right) \tag{2.51}$$

where $\tau = bt$ and

$$a_1\psi\left(\tau/b\right) = \begin{cases} A, & \text{for } \tau_n < \tau < \tau_m \\ -B, & \text{for } \tau_m < \tau < \tau_{n+1} \end{cases}. \tag{2.52}$$

The regeneration points, τ_n and τ_m, are defined through two dimensionless time variables, T_1 and T_2,

$$\tau_n = n\left(T_1 + T_2\right); \qquad \tau_m = \tau_n + T_1. \tag{2.53}$$

As special cases, this form of the equations includes the symmetric pulse, $A = B$ and $T_1 = T_2 = T/2$, and a train of a delta-function pulses, $B = 0$, $A \to \infty$ and $T_1 \to 0$ with $AT_1 = 1$. The latter case has been considered in a more general setting [56].

For a constant value of $a_1\psi\left(\tau/b\right)$, Eq. (2.51) reduces to Eq. (2.1) (with a replaced by $a + A$ and $a - B$) with known solutions (2.2) and (2.3). Therefore, one has to write the solutions of (2.51) in the two regimes given in (2.52), and then match the solutions at the boundaries of the appropriate regimes. The details of the calculation can be found in [57]. The final result of these calculations is the transcendental equation which defines the boundaries of the Shapiro steps.

Chapter 3

Underdamped Pendulum

3.1 Pendulum with constant torque

The pendulum equation with constant torque is obtained by adding a constant torque a to Eq. (1.2),

$$\frac{d^2\phi}{dt^2} + \gamma\frac{d\phi}{dt} + b\sin\phi = a. \tag{3.1}$$

This equation cannot be solved analytically, and we shall describe the results of numerical calculations [58], [59].

For $a = 0$, the pendulum hangs in the downward position, $\phi = 0$. With increasing a, the equilibrium state is tilted to $\phi = \sin^{-1}(a/b)$, approaching $\phi = \pi/2$ at $a = b$. For $a > b$, the pendulum starts to rotate, corresponding to the loss of equilibrium. Note that this result corresponds to Eqs. (2.1) and (2.4) for the overdamped pendulum. The type of rotation depends on the damping. If γ is small, the pendulum rotates with a finite period. For strong damping and close to the threshold, $a - b$ of order the small parameter ε, and the period of rotation scales as $\varepsilon^{-1/2}$.

As a decreases from $a > b$ to $a < b$, the pendulum continues to rotate even for $a < b$, which is the manifestation of hysteresis. For $a \leq b$, if we stop the rotating pendulum by hand and put it close to its equilibrium state, the pendulum will remain in equilibrium, showing bistability between the rotating periodic regime and stationary equilibrium. Another peculiarity occurs when the pendulum is perturbed from the equilibrium position $\phi_1 = \sin^{-1}(a_1/b)$ by releasing it without initial velocity. The reaction of the pendulum then depends of the direction of the perturbation. If the perturbation is directed toward the vertical direction, the pendulum returns to equilibrium. But if the perturbation is large enough and directed in the opposite direction, the pendulum moves in the direction of the perturbation.

Equation (3.1) can be rewritten as two first-order differential equations by introducing the dimensionless angular velocity

$$z = \frac{\gamma}{b}\frac{d\phi}{dt},\tag{3.2}$$

implying $dz/dt = (bz/\gamma)(dz/d\phi)$. Then, one can rewrite Eq. (3.1) in the form

$$\frac{bz}{\gamma^2}\frac{dz}{d\phi} + z + \sin\phi = \frac{a}{b}\tag{3.3}$$

where $a/b \equiv \bar{a}$ is the dimensionless torque.

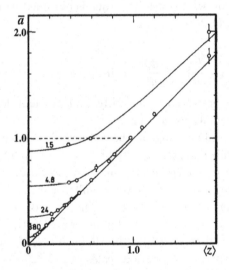

Fig. 3.1   Normalized average angular velocity $\langle z \rangle$ as a function of normalized torque $\bar{a}$ for different values of $\beta = b/\gamma^2$. The curves have been drawn to guide the eye.

Solutions of Eqs. (3.2)-(3.3) plotted in the $(\langle z \rangle, \bar{a})$ plane are shown in Fig. 3.1. It is seen that two solutions exist for the average angular velocity $\langle z \rangle$ in the interval $1 > \bar{a} > \bar{a}_{cr}$, where $\bar{a}_{cr}$ depends on the parameter b/γ^2. The smaller this parameter, the larger the bias torque $\bar{a}$ that one needs to start the motion, with some threshold value of $\langle z \rangle$. On the other hand, by decreasing the torque $\bar{a}$, the pendulum will continue to rotate until the torque reaches the value $\bar{a}_{cr}$, at which point the pendulum will come to rest after performing damped oscillations around the equilibrium position $\phi = \sin^{-1}(\bar{a}_{cr}/b)$.

3.2 Pendulum with multiplicative noise

If the rod of a pendulum performs random vibrations, one has to add the external multiplicative noise $\xi(t)$ to the equation of motion (1.1),

$$\frac{d^2\phi}{dt^2} + \left[\omega_0^2 + \xi(t)\right]\sin\phi = 0. \tag{3.4}$$

Equation (3.4) can be rewritten as two first-order stochastic differential equations

$$\frac{d\phi}{dt} = \Omega, \tag{3.5}$$

$$\frac{d\Omega}{dt} = -\left[\omega_0^2 + \xi(t)\right]\sin\phi \tag{3.6}$$

with the energy of the system given by Eq. (1.4). The analysis of Eqs. (3.5)-(3.6) is quite different for white noise and for colored noise [60]. The Fokker-Planck equation associated with the Langevin Eqs. (3.5)-(3.6) has the following form,

$$\frac{\partial P}{\partial t} = -\frac{\partial}{\partial\phi}(\Omega P) + \frac{\partial}{\partial\Omega}\left[\omega_0^2\sin(\phi)P\right] + \frac{D}{2}\sin^2\phi\frac{\partial^2 P}{\partial\Omega^2}. \tag{3.7}$$

To use the averaging technique, one notes that according to Eq. (3.5), the variable ϕ varies rapidly compared with Ω. Therefore, in the long-time limit, one can assume [61] that the angle ϕ is uniformly distributed over $(0, 2\pi)$. Hence, one can average Eq. (3.7) over ϕ, which gives a Gaussian distribution for the marginal distribution function $P_1(\Omega)$,

$$P_1(\Omega) = \frac{1}{\sqrt{\pi Dt}}\exp\left(-\frac{\Omega^2}{Dt}\right). \tag{3.8}$$

Accordingly, one obtains for the energy $E \approx \Omega^2/2$,

$$P_1(E) = \sqrt{\frac{2}{\pi Dt}}E^{-1/2}\exp\left(-\frac{2E}{Dt}\right) \tag{3.9}$$

and from (3.9), it follows that

$$\langle E \rangle = \frac{D}{4} t. \tag{3.10}$$

For colored noise, one cannot write the exact Fockker-Planck equation. However, if one assumes a power-law dependence of ϕ as a function of time t, the self-consistent estimate gives [60],

$$\langle E \rangle \approx \sqrt{t}. \tag{3.11}$$

These heusteric arguments are confirmed [60] by more rigorous analysis and by numerical calculations.

3.3 Pendulum with additive noise

3.3.1 *Damped pendulum subject to additive noise*

The equation of motion for the angle ϕ of a damped pendulum subject to a constant torque a, a random torque $\xi(t)$, and a periodic force $f \sin(\omega t)$ acting on the bob (Fig. 3.2), has the following form,

$$\frac{d^2\phi}{dt^2} + \gamma \frac{d\phi}{dt} + b \sin \phi = a + f \sin(\omega t) + \xi(t). \tag{3.12}$$

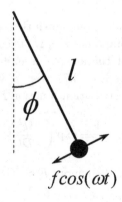

Fig. 3.2   Pendulum with an external periodic force acting on the bob.

In this section we consider different simplified forms of Eq. (3.12). We begin with the following form,

$$\frac{d^2\phi}{dt^2} + \gamma\frac{d\phi}{dt} + b\sin\phi = \xi(t). \tag{3.13}$$

According to the fluctuation-dissipation theorem for a stationary state, a gain of energy entering the system is exactly compensated by the energy loss to the reservoir, which gives

$$\langle \xi(t)\,\xi(0) \rangle = 2\gamma\kappa T \tag{3.14}$$

where κ is the Boltzmann constant. The velocity-velocity autocorrelation function defines the frequency-dependent mobility $\mu(\omega)$,

$$\mu(\omega, T) = \frac{1}{\kappa T}\int\limits_0^\infty dt\,\left\langle \frac{d\phi}{dt}(t)\,\frac{d\phi}{dt}0)\right\rangle \exp(i\omega t). \tag{3.15}$$

For special cases, one can obtain analytical results for $\mu(0, T)$. For the non-damped case ($\gamma = 0$) [62],

$$\mu(0, T) = \pi(2\pi\kappa T)^{-1/2}\frac{I_0(v) + I_1(v)}{I_0^2(v)}\exp(-v) \tag{3.16}$$

where $I_n(v)$ denotes the Bessel function of n-th order. In the limit of large damping [35] and neglecting the second derivative in (3.13),

$$\mu(0, T) = \frac{1}{\gamma\,[I_0^2(v)]}. \tag{3.17}$$

The method of continued fraction [14] yields an approximate expression for the mobility [63],

$$\mu(\omega) \approx \frac{1}{-i\omega + \gamma + ibI_1(0)/\omega I_0(v)} \tag{3.18}$$

which is in good agreement with numerical calculations [63].

3.3.2 *Damped pendulum subject to constant torque and noise*

The analytical solution of the simple mathematical pendulum, Eq. (3.12) with $\gamma = f = \xi = 0$, has been considered in Sec. 1.1. If one adds damping while keeping $f = 0$, Eq. (3.12) takes the form,

$$\frac{d^2\phi}{dt^2} + \gamma\frac{d\phi}{dt} = a - b\sin\phi + \xi(t) \qquad (3.19)$$

which describes a damped pendulum subject to the potential

$$U(\phi) = -a\phi + b(1 - \cos\phi). \qquad (3.20)$$

In Sec. 2.2, the analysis of overdamped deterministic motion (neglecting the second derivative in Eq. (3.12) and the random force $\xi(t)$) showed that there are two types of solutions: locked-on for $a < b$, and running for $a > b$. A similar situation occurs [64] for Eq. (3.19). The system switches randomly between a locked state with zero average velocity, $\langle d\phi/dt \rangle = 0$, and a running state with $\langle d\phi/dt \rangle \neq 0$. For $\gamma << \sqrt{b}$, the transition from the locked state to the running state occurs when the value of a approaches b, whereas the opposite transition takes place when the value of a decreases to $4\gamma\sqrt{b}/\pi$. For large values of η, say, $\gamma > \pi\sqrt{b}/4$, we recover the overdamped case, and the two boundary values coincide, coming back to the criterion $a \gtrless b$. For weak noise, the stationary dynamics of $\phi(t)$ is controlled by the critical value of $a_{cr} = (2 + \sqrt{2})\gamma\sqrt{b}$. For $a < a_{cr}$, a system is trapped in small oscillations near the downward position, whereas for $a > a_{cr}$, the pendulum performs rotations.

For the running solutions, the average angular velocity $\langle d\phi/dt \rangle$ will be non-zero. This velocity and the time evolution of the mean square displacement of the rotation angle $\langle \phi^2 \rangle$ are the characteristics of the pendulum motion. When $\langle \phi^2 \rangle$ is proportional to t^μ with $\mu = 1$, the process is called normal diffusion, whereas for $\mu \neq 1$, the diffusion is anomalous, and is called super- or sub-diffusion for $\mu > 1$ and $\mu < 1$, respectively. An exact analytical solution can be only obtained for the simple pendulum described by Eq. (1.1). In all other cases, one has to rely on numerical solutions. Note that that in the absence of any stochastic force ($\xi = 0$ in Eq. (3.12)), "deterministic" diffusion occurs.

An unexpected phenomenon has been found [65] for intermediate values of the parameter a. This is precisely the region which is very important for experimentalists, but is rarely discussed by theorists who study (analytically) the asymptotic time regime and (numerically) the initial time regime. In addition to the characteristic value a_{cr} of the parameter a described above, which defines the boundary between locked and running trajectories, there is another characteristic value of parameter a, well below a_{cr}, where the system alternates between the locked and running trajectories.

The new regime lies [65] in the intermediate region between these two characteristic values of a, in which the system moves coherently with constant velocity (in spite of the noise!).

3.4 Periodically driven pendulum

For the simplified case, $\gamma = a = \xi = 0$ in (3.12), detailed numerical analysis has been performed [66] for $f = 1.2$ and two different values of ω (0.1 and 0.8). The solution is quite different for these two cases: in the former case, the diffusion is normal, whereas in the latter case, the diffusion is anomalous with $\mu = 1.6$. We will show in Part 4 that this difference stems from the different structure of the Poincare map, which only has regular regions for $\omega = 0.1$, but has both regular and chaotic regions for $\omega = 0.8$.

When damping is included, the equation of motion (3.12) takes the following form,

$$\frac{d^2\phi}{dt^2} + \gamma\frac{d\phi}{dt} + \frac{g}{l}\sin\phi = f\sin(\omega t). \tag{3.21}$$

Analysis of this equation has been performed [66] for $g/l = 1$, $f/m = 1.2$, $\gamma = 0.2$ and different values of ω. Chaotic behavior starts at the accumulation point of period-doubling bifurcations, $\omega_1 = 0.47315$. Depending on the value of ω, the influence of the regions with normal diffusion and chaotic bursts is different. In the region to the left of ω_1, ballistic motion (normal diffusion) is dominant with the burst regions contributing only minimally. The diffusion coefficient D exhibits a power-law divergence $D \approx |\omega - \omega_1|^{-\alpha}$, with the exponent α close to $\frac{1}{2}$. For another type of bifurcation, called crisis induced intermittency, analysis of equation

$$\frac{d^2\phi}{dt^2} + \gamma\frac{d\phi}{dt} + \sin\phi = f\cos(\omega t) \tag{3.22}$$

shows [67] that for $f < f_c$ (where f_c is the bifurcation point), there exist two attractors which merge for $f \geq f_c$, and form a single large attractor with chaotic switching between these two attractors. The mean time between switching is a power-law function of $f - f_c$, with the exponent given in terms of the expanding and contracting eigenvalues [67]. For $\omega = 1.0$, $\gamma = 0.22$, $f_c = 2.646442$, the diffusion coefficient was calculated [66] in the regime $f \in [2.64653, 2.69]$ and is described by the power law, $D \approx (f - f_c)^{0.699}$.

Extensive studies of diffusion in this case have been performed for Eq. (3.22) which can be rewritten in dimensionless form [68; 69],

$$\frac{d^2\phi}{dt^2} + \frac{1}{Q}\frac{d\phi}{dt} + \sin\phi = \epsilon\sin(\Omega t) \qquad (3.23)$$

and contains three independent dimensionless parameters: $Q = ml\sqrt{gl/\gamma}$, $\epsilon = f/mgl$ and $\Omega = \omega\sqrt{l/g}$. In [69], two variables were fixed, $\Omega = 0.6$ and $Q = 5$, while the amplitude ϵ of the external periodic force varied through the interval $(0.6 - 1.0)$. Many different solutions have been found - periodic, non-periodic and chaotic - and special attention was paid to the symmetry properties of these solutions.

In order to study diffusion, two parameters are fixed, $\epsilon = 0.78$ and $\Omega = 0.62$, while the parameter Q was allowed to take 200 different values in the interval $3 < Q < 7$. The authors argue that their results are not dependent on the special values assumed for the parameters. Based on the exponential dependence of the chaotic states on the initial conditions, numerical calculations have been performed for ensemble-averaged $\langle\phi^2\rangle$, where the averaging was taking over a large grid of initial conditions centered around the origin, $\phi = d\phi/dt = 0$. The numerics prove that diffusion remains normal, with the diffusion coefficient depending essentially on Q and is considerably enhanced for Q corresponding to the edges of periodic windows in the bifurcation diagram. The periodic states located within the windows do not diffuse, which corresponds to $D = 0$. However, even for some running solutions which expand in phase space, the motion is ballistic, i.e., $\langle\phi^2\rangle$ is a quadratic rather than a linear function of time. These results are modified if one adds quenched disorder [70] to Eq. (3.23),

$$\frac{d^2\phi}{dt^2} + \frac{1}{Q}\frac{d\phi}{dt} + \sin\phi = \epsilon\sin(\Omega t) + \alpha\eta(x) \qquad (3.24)$$

where $\eta(x) \in [-1, 1]$ are independent, uniformly distributed random variables, and α gives the amount of quenched disorder. The presence of quenched disorder destroys the periodicity and symmetry of Eq. (3.23), and modifies the results. Whereas there were only running solutions for $\alpha = 0$, both locked and running solutions appear for $\alpha \neq 0$, depending on initial conditions. Although the diffusion remains normal, $\lim_{t\to\infty}\left(\langle x^2\rangle/2t\right) = D(\alpha)$, ie., the diffusion coefficient D depends on α, so that for $\alpha < 0.1$, $D(\alpha) > D(0)$, whereas for $\alpha > 0.1$, $D(\alpha) < D(0)$. The latter result is due to the increase of the locked trajectories when α is increased.

In addition to the diffusion considered above and discussed in Part 4 for chaotic motion, some further interesting results have been obtained [71]. These workers analyzed Eq. (3.22) for the case in which the frequency ω of an external field is larger than the eigenfrequency of the pendulum ω_0, which is equal to unity in Eq. (3.22). For fixed values of parameters, $\gamma = 0.2$ and $\omega = 2$, they obtained numerical and approximate analytical solutions of Eq. (3.22) for different values of the amplitude f of the external field. They found that no chaotic regime appears far from the resonance condition $\omega \simeq \omega_0$, which is replaced by oscillations around the upward position (see Part 5).

The conventional stationary solution of Eq. (3.22) takes the following form,

$$\phi(t) = A_1 \cos\theta_1 + A_3 \cos\theta_3 + A_5 \cos\theta_5 + \cdots ; \quad \theta_m = m\omega t - \alpha_m; \quad 0 < \alpha_m < \pi. \tag{3.25}$$

The coefficient A_1 increases with f, and when f exceeds $f_c \approx 8.77$, A_1 exceeds $A_c \approx 2.425$, which is close to the first zero (2.405) of the Bessel function $J_0(A_1)$. Symmetry breaking then occurs with respect to $\phi \to -\phi$, and the stationary solution (3.25) is replaced by the symmetry-breaking solution

$$\phi(t) = A_0 + A_1 \cos\theta_1 + A_2 \cos\theta_2 + A_3 \cos\theta_3 + \cdots . \tag{3.26}$$

At this stage, deterministic chaos-like phenomena occur. A change in the initial condition $\phi(t = 0)$ as small as 10^{-7} leads to a change of A_0 in (3.26) by a significant amount. However, such sensitivity to the initial conditions exists only for some particular values of f, in contrast to deterministic chaos where sensitivity exists almost everywhere throughout a definite region of parameter space. On further increase of f, $|A_0|$ reaches π at $f = 10.58$, where the pendulum oscillates around the upward position. As f continuous to increase, A_1 increases while $|A_0|$ remains at π until A_1 reaches approximately the second zero (5.43) of $J_0(A_1)$. Then, $|A_0|$ decreases with A_1 until A_0 reaches zero and the pendulum returns to the stationary regime (3.25). After this, A_1 again increases until it reaches the third zero (8.65) of $J_0(A_1)$, and so on. In the regime $f \in (3.16, 5.97)$, there appears a period-triple state, approximately described by $\phi = A_1 \cos(\omega t - \alpha_1) + A_{1/3} \cos\left[(\omega/3)t - \alpha_{1/3}\right]$, and in the regime $f \in (3.77, 11.40)$, there appears a sinusoidally modulated rotational state approximately described by $\phi = A_0 + A_1 \cos(\omega t - \alpha_1) \pm \omega t$. The two signs in the last formula correspond to the two directions of rotation. Note that these two modulated states, period-triple states, along with the single-period states (3.25) and (3.26), can occur for the same value of f, with the initial conditions determining which state is actually excited. All the above results obtained from numerical calculations are in good agreement with the analytic calculations performed [71] in the framework of perturbation theory.

3.5 Damped pendulum subject to constant torque, periodic force and noise

The addition of a constant term a to the right hand side of Eq. (3.23) leads to the loss of periodicity. As a increases, the system alternatively becomes chaotic and periodic. One of the two intermittent regions described in the previous section is destroyed by a small bias a ($a = 0.0005$ for $\omega = 0.47311$, $\gamma = 0.2$, and $f = 1.2$), and the motion becomes nearly ballistic, in contrast to the case without bias, for which two symmetric intermittent regions lead to normal diffusion [66]. These counterintuitive results have been obtained [72] for the general case, when, in addition to a constant term a, thermal noise is also included in Eq. (3.16), which returns us to the original Eq. (3.12). Numerical calculations have been performed [72] for the asymptotic mean velocity $\langle\langle d\phi/dt\rangle\rangle$, which is defined as the average of the angular velocity over time and thermal fluctuations. For values of parameter a lying in the interval $(-a_c, a_c)$, the velocity is oriented in the opposite direction from that of the driving force a, thereby displaying "absolute negative mobility". This phenomenon, obtained for a non-equilibrium system, is opposite to that obtained for an equilibrium system for which the response is always in the same direction as the applied force. The value of the parameter a_c depends of the values of other parameters entering Eq. (3.12).

A pendulum subject to both additive noise and multiplicative noise has recently been considered [73]. The equation of motion is obtained by adding multiplicative noise to Eq. (3.12),

$$\frac{d^2\phi}{dx^2} + \gamma\frac{d\phi}{dt} + [a + \sigma\xi(t)]\sin\phi = a + f\sin(\Omega t) + \eta(t) \qquad (3.27)$$

where $\xi(t)$ and $\eta(t)$ assumed to be delta-correlated Orenstein-Uhlenbeck noises,

$$\langle \xi(t)\,\xi(t_1)\rangle = E^2 \exp\left(-\lambda\,|t - t_1|\right),$$

$$\langle \eta(t)\,\eta(t_1)\rangle = sE^2 \exp\left(-\lambda\,|t - t_1|\right), \qquad (3.28)$$

$$\langle \xi(t)\,\eta(t_1)\rangle = 2D\,\delta(t - t_1).$$

Numerical simulation shows [73] that if the product $D\lambda$ is constant, the system is either in a locked state or in a running state as λ changes. For small λ, for $t \to \infty$, the system is randomly located in one of two stable states, $\phi = 0$ or $\phi = -\pi$. For large λ, the system is always in the state $\phi = 0$, and as λ increases, ϕ undergoes the transition monostability-bistability-monostability. The coefficient σ of multiplicative noise in (3.27) simulates the time evolution of ϕ: the running solutions start from $|\sigma| > 0$, and ϕ turns over counterclockwise at $\sigma = -20$. Furthermore, the larger the value of σ, the faster ϕ turns over. This means that the strength of multiplicative noise controls the rotation direction of ϕ without any external torque. The bistability regime exists for any value of additive noise, whose strength serves only to change the turnover direction and speed of ϕ.

3.6 Pendulum with oscillating suspension point

3.6.1 *Vertical oscillations*

Thus far, we have considered the pendulum as a massless rod of length l with a point mass (bob) m at its end (Fig. 1.1), and our analysis dealt only with a pendulum whose suspension point is at rest. When the suspension point performs vertical oscillations $u(t) = a\sin(\omega t)$ (Fig. 3.3), the system becomes non-inertial, and it is convenient to use a non-inertial frame of reference fixed to the moving axis. Newton's law must then be "modified" by the addition of the force of inertia $-m\,d^2u(t)/dt^2$, which produces a torque $-ml\sin\phi\; d^2u/dt^2$. This means that for a pendulum with an oscillating suspension point, one replaces gravity g by $g - d^2u/dt^2$, which gives

$$\frac{d^2\phi}{dt^2} + \frac{1}{l}\left[g + a\omega^2 \sin(\omega t)\right]\sin\phi = 0. \qquad (3.29)$$

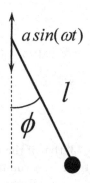

Fig. 3.3   Pendulum with vertically oscillating suspension point.

In terms of the dimensionless time variable $\tau = \omega t$, Eq. (3.29) becomes

$$\frac{d^2\phi}{d\tau^2} + [\alpha + \beta \sin(\tau)] \sin \phi = 0; \qquad \alpha = \frac{g}{l\omega^2}; \qquad \beta = \frac{a}{l}. \qquad (3.30)$$

Equation (3.30) cannot be solved analytically. Numerical calculations show [74] the different dynamic modes. For example, for $\beta = -0.1$, the downward position is stable for $\alpha = 0.5$ and unstable for $\alpha = 0.25$. The upward position, which we will consider in detail in Part 5, is stable for $\alpha = -0.1$, $\beta = -0.545$, but becomes unstable for $\alpha = -0.2$, $\beta = -0.1$.

Introducing damping into (3.30) leads to

$$\frac{d^2\phi}{d\tau^2} + \gamma \frac{d\phi}{d\tau} + [\alpha + \beta \sin(\tau)] \sin \phi = 0. \qquad (3.31)$$

Depending on the initial conditions, for the parameters $\alpha = 0.5$, $\beta = -0.1$, and $\gamma = 0.03$, there are four different asymptotic $(t \to \infty)$ solutions of Eq. (3.31): the stationary state at $\phi = d\phi/dt = 0$, the lock-in oscillating solutions near this point, and the clockwise and counterclockwise running solutions. The types of solutions are different for other values of the parameters in (3.31). For example, for $\alpha = 0.02$, $\beta = -0.35$, $\gamma = 0.03$, there are five types of solutions (the upward stationary solutions in addition to the four described above), and for $\alpha = -0.1$, $\beta = -0.545$, $\gamma = 0.08$, there are three types of the asymptotic solutions (the stationary state at $\phi = \pi$, $d\phi/dt = 0$ and the clockwise and couterclockwise running solutions). Comprehensive numerical analysis shows [74] which initial values of ϕ and $d\phi/dt$ lead to each type of asymptotic solution.

For analyzing Eq. (3.31), it is useful [75] to consider the so-called rotational number $\langle v \rangle / \omega$, defined as

$$\frac{\langle v \rangle}{\omega} = \lim_{\tau_2 \to \infty} \frac{1}{\omega \left(\tau_2 - \tau\right)} \int_{\tau_1}^{\tau_2} \frac{d\phi}{dt} d\tau \qquad (3.32)$$

where τ_1 is the time interval necessary for the system to reach steady-state for large enough τ_2. Numerical calculation of the rotational number as a function of the amplitude β of the external field shows [75] that for $\beta \approx 0.47$, $\langle v \rangle / \omega$ remains constant. However, for higher β, $\langle v \rangle / \omega = \pm 1$ for running solutions (rotations) and equals zero for locked (oscillating) solutions. These two modes, locked and running, are clearly decoupled.

Of great importance for many applications are the values of the parameters that determine the boundaries between locked and running solutions ("escape parameter region"). This problem is analogous to the escape of a particle from a potential well. To solve this problem for the parametrically excited damped pendulum that is symmetric around $\phi = 0$, as described by Eq. (3.31) with $\alpha = 1$, one uses the harmonic balance method [76]. One obtains the following equation defining the transition from the locked to the running solutions in the $\omega - \beta$ plane,

$$\left(\frac{\omega^2}{4} - 1\right)^2 + \frac{\gamma^2 \omega^4}{4} - \frac{\beta^2 \omega^4}{4} = 0. \qquad (3.33)$$

An unconventional approach has recently been taken [77] to the analytical solution of the problem of the locked-running transition by transforming the differential equation (3.31) into an integral equation. Equation (3.31) can be rewritten as

$$\frac{d^2 \phi}{dt^2} + \gamma \frac{d\phi}{dt} + [1 + f \cos\left(\Omega t\right)] \sin \phi = 0. \qquad (3.34)$$

By multiplying Eq. (3.34) by $d\phi/dt$, one obtains the law of the conservation of energy,

$$\frac{d}{dt}\left[\frac{1}{2}\left(\frac{d\phi}{dt}\right)^2 - \cos\phi\right] = -\gamma \left(\frac{d\phi}{dt}\right)^2 - f \frac{d\phi}{dt} \cos\left(\Omega t\right) \sin\phi. \qquad (3.35)$$

Integrating this equation yields

$$\left(\frac{d\phi}{dt}\right)^2 - 2\cos\phi - \left(\frac{d\phi}{dt}\right)_0^2 + 2\cos\phi_0$$

$$= -2\gamma \int_{t_0}^t \frac{d\phi(r)}{dr}\frac{d\phi}{dr}dr - 2f \int_{t_0}^t \cos(\Omega r)\sin[\phi(r)]\frac{d\phi}{dr}dr \qquad (3.36)$$

where ϕ_0 and $(d\phi/dt)_0$ are the initial conditions at time t_0. For rotations, $d\phi/dt \neq 0$, and one can find the inverse function $t = t(\phi)$. For the function $d\phi/dt = \psi[\phi(t)]$, the inverse function is

$$t = t_0 + \int_{t_0}^t \frac{ds}{\psi(s)}. \qquad (3.37)$$

One then rewrites Eq. (3.36) as

$$\psi(\phi)^2 - 2\cos\phi - \left(\frac{d\phi}{dt}\right)_0^2 + 2\cos\phi_0$$

$$= -2\gamma \int_{\phi_0}^\phi \psi(s)\,ds - 2f \int_{\phi_0}^\phi \cos\left(\Omega t_0 + \Omega \int_{\phi_0}^z ds/\psi(s)\right)\sin z\,dz. \qquad (3.38)$$

For rotations with period T,

$$\phi(T) = \phi(t_0 + T) = \phi_0 + 2\pi; \quad \frac{d\phi}{dt}(T) = \frac{d\phi}{dt}(t_0 + T) = \left(\frac{d\phi}{dt}\right)_0. \qquad (3.39)$$

Using Eq. (3.39), one can rewrite Eq. (3.36) for $\phi = \phi_0 + 2\pi$ in the following form,

$$\gamma A + f\cos(\Omega t_0) B - f\sin(\Omega t_0) C = 0, \qquad (3.40)$$

where

$$A = \int_{\phi_0}^{\phi_0+2\pi} \psi(s)\,ds, \quad B = \int_{\phi_0}^{\phi_0+2\pi} \cos\left(\Omega \int_{\phi_0}^z ds/\psi(s)\right)\sin z\,dz,$$

$$(3.41)$$

$$C = \int_{\phi_0}^{\phi_0+2\pi} \sin\left(\Omega \int_{\phi_0}^z \frac{ds}{\psi(s)}\right)\sin z\,dz.$$

Defining the angle α by

$$\sin\alpha = \frac{B}{\sqrt{B^2+C^2}}; \quad \cos\alpha = \frac{C}{\sqrt{B^2+C^2}}, \qquad (3.42)$$

Eq. (3.40) becomes

$$\sin\left(\Omega t_0 - \alpha\right) = \frac{\gamma}{f}\frac{A}{\sqrt{B^2 + C^2}}. \tag{3.43}$$

One concludes from (3.43) that a periodic solution exists if and only if the amplitude f of the external force satisfies

$$f \geq f_{cr} = \frac{\gamma A}{\sqrt{B^2 + C^2}}. \tag{3.44}$$

Equation (3.44) was obtained by an exact analytical calculation for given function $\psi\left(\phi\right)$ which defines the trajectory $d\phi/dt = \psi\left(\phi\right)$ in phase space. This approach allows one to build a convenient form of perturbation theory [77].

3.6.2 *Horizontal oscillations*

When the suspension point executes periodic horizontal oscillations, $f\cos\left(\omega t\right)$ (Fig. 3.4), the coordinates x, z of the bob become

$$z = l\cos\phi; \qquad x = l\sin\phi + f\cos\omega t \tag{3.45}$$

which leads to the following kinetic energy T and potential energy U,

$$T = \frac{1}{2}m\left[\left(dx/dt\right)^2 + \left(dz/dt\right)^2\right]$$

$$= \frac{1}{2}m[l^2\left(d\phi/dt\right)^2 - 2l\omega f\sin\left(\omega t\right)\cos\phi\left(d\phi/dt\right) + f^2\omega^2\sin^2\left(\omega t\right)]; \tag{3.46}$$

$$U = -mgl\cos\phi.$$

Substituting Eq. (3.46) into the Lagrange function $L = T - U$, and using the Lagrange equation, leads to the following equation of motion,

$$\frac{d^2\phi}{dt^2} + \frac{k}{l}\frac{d\phi}{dt} + \omega_0^2\sin\phi - \frac{f}{l}\omega^2\cos\left(\omega t\right)\cos\phi = 0 \tag{3.47}$$

where $\omega_0^2 = g/l$, and the damping term proportional to $d\phi/dt$ has been added.

3.6.3 *Pendulum with parametric damping*

The external force acting on a pendulum may be introduced in several different ways. In addition to a periodic torque and the oscillation of the

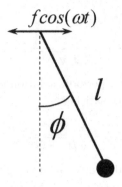

Fig. 3.4 Pendulum with horizontally oscillating suspension point.

suspension point, one can add a time-dependent term to the damping,

$$\frac{d^2\phi}{d\tau^2} + [A + B\sin(\omega\tau)]\frac{d\phi}{d\tau} + \omega_0^2\sin\phi = 0 \tag{3.48}$$

which takes the form

$$\frac{d^2\phi}{dt^2} + \frac{1}{Q}[1 + f\sin(\Omega t)]\frac{d\phi}{dt} + \sin\phi = 0 \tag{3.49}$$

where $\Omega = \omega/\omega_0$ and $t = \omega_0\tau$ is the dimensionless time.

The numerical solution of Eq. (3.49) has been obtained [78] for $Q = 18.33$, using the amplitude f and the frequency ratio Ω as control parameters. In contrast to Eq. (3.31), the point $\phi = d\phi/dt = 0$ satisfies Eq. (3.49). For $\Omega \gtrsim 2$ and arbitrary f, this point is the accumulation point of all damping solutions of Eq. (3.49). For smaller Ω, the solutions are periodic, and for intermediate values of Ω, $[\Omega \in (1.5 - 2.0)]$, a long chaotic transient precedes the approach to the point $\phi = d\phi/dt = 0$. Moreover, there are some values of f and Ω for which other attractors appear, dependent on the initial conditions, or those for which the point $\phi = d\phi/dt = 0$ becomes unstable. Finally, for a small regime of parameters f and Ω, the motion is chaotic. A detailed analysis of the chaotic solutions that appear for $\Omega \approx 2$ has been performed [79]. It was found that the minimum amplitude f needed to drive the system into chaos occurs at the frequency $\Omega = 1.66$, whereas the minimum value of f needed to destabilize the stationary solutions occurs for the following values of parameters: $f = 2$, $\Omega = 2$. It is remarkable that these properties are insensitive to the value of the damping. All these theoretical results have been confirmed on a specially constructed experimental setup [78], [80].

Apart from the numerical calculation, the approximate analytical treatment has been performed based on the Floquet theorem [81]. Since the stability of the fixed point $\phi = 0$, $d\phi/dt = 0$, and the dynamics near this point were their main interest, the authors [81] analyzed the linearized equation of motion. By using the transformation

$$\tau = \frac{\Omega t}{2}; \qquad \omega = \frac{\Omega}{2}; \qquad q = \frac{\Omega}{2Q}; \qquad \varepsilon = \frac{f}{2}, \tag{3.50}$$

the linearized Eq. (3.49) can be rewritten as

$$\omega^2 \frac{d^2\phi}{d\tau^2} + q\left[1 + \varepsilon \sin\left(2\tau\right)\right] \frac{d\phi}{d\tau} + \phi = 0. \tag{3.51}$$

According to the Floquet theorem, the solution of Eq. (3.51) has the form,

$$\phi = A_0 + \exp\left(\mu t\right) \sum_{n=1}^{\infty} \left[A_n \cos\left(nt\right) + B_n \sin\left(nt\right)\right]. \tag{3.52}$$

The boundary of stability is determined by the existence of a periodic solutions for ϕ at $\mu = 0$. Inserting (3.52) with $\mu = 0$ into (3.51) and collecting the coefficients of the sine and cosine terms leads to

$$\begin{aligned}
\left(1 - n^2\omega^2\right) A_n + nqB_n + q\varepsilon\left[(n-2) A_{n-2} - (n+2) A_{n+2}\right] &= 0, \\
\left(1 - n^2\omega^2\right) B_n - nqA_n + q\varepsilon\left[(n-2) B_{n-2} - (n+2) B_{n+2}\right] &= 0.
\end{aligned} \tag{3.53}$$

It follows from (3.53) that these equations separate into two classes, odd n and even n, whereas nontrivial solutions exist only for odd n. The solutions for the latter depend on the truncation order. For truncation at $n = 1$, there are two homogeneous equations for A_1 and B_1. Nontrivial solutions exist when the determinant of these equations vanishes, which gives for the boundary of stability,

$$\varepsilon_{cr}^1 = q^{-1}\sqrt{\left(1 - \omega^2\right)^2 + q^2}. \tag{3.54}$$

For $Q = 18.33$ and $\Omega = 1.5$, the amplitude ε_{cr} of the external field that imposes the onset of instability is $\varepsilon_{cr}^1 = 21.48$ for truncation at $n = 1$. Analogously, for truncation at higher odd n, one obtains [81] $\varepsilon_{cr}^3 = 17.08$, $\varepsilon_{cr}^5 \approx \varepsilon_{cr}^7 = 17.18$. This result is a slight improvement over the value $\varepsilon_{cr} = 17.2$ obtained [78] by numerical calculation.

Analysis of the dynamics shows [81] that for $\varepsilon > \varepsilon_{cr}$, the Floquet multiplier μ is proportional to the distance to the boundary of stability, $\mu \approx \varepsilon - \varepsilon_{cr}$.

3.7 Spring pendulum

A spring pendulum is obtained from a rigid pendulum by inserting into the rod a spring of stiffness constant κ and unstretched length l_0 (Fig. 3.5). Since the system has a period of 2π in the angle ϕ, there are two fixed points, $\phi = 0$ and $\phi = \pi$. One chooses the origin at the suspension point with the z-axis pointing upward, as shown in Fig. 3.5.

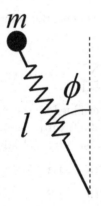

Fig. 3.5   Spring pendulum.

In the downward equilibrium position, $\phi = \pi$, at which point the elastic stress of the spring is balanced by gravity acting by the bob,

$$\kappa\left(l - l_0\right) = mg, \qquad \text{or} \quad \frac{l}{l_0} - 1 = \frac{\omega_0^2}{\omega_s^2} \tag{3.55}$$

where $\omega_0^2 = g/l_0$ and $\omega_s^2 = \kappa/m$. Then,

$$l = l_0\left(1 + \frac{\omega_0^2}{\omega_s^2}\right). \tag{3.56}$$

In our coordinate system, the downward position of the bob is defined by $x = 0$ and $z = -l$,

$$z = -l = -l_0 \left(1 + \frac{\omega_0^2}{\omega_s^2}\right). \tag{3.57}$$

Let the length of the pendulum in the vertical upward position be $\hat{l}$. In equilibrium, the elastic stress and gravity act in the same direction,

$$\kappa \left(\hat{l} - l_0\right) + mg = 0 \tag{3.58}$$

or

$$\hat{l} = l_0 \left(1 - \frac{\omega_0^2}{\omega_s^2}\right). \tag{3.59}$$

In the vertical position, the coordinates of the bob are $x = 0$, $z = l_0 \left(1 - \omega_0^2/\omega_s^2\right)$. It is clear from Eq. (3.59) that $\omega_s > \omega_0$.

The Lagrangian of the system has the following form,

$$L \equiv T - U = \frac{m}{2}\left[\left(\frac{dx}{dt}\right)^2 + \left(\frac{dz}{dt}\right)^2\right] + mgz - \frac{\kappa}{2}(R - l_0)^2 \tag{3.60}$$

where $R = \sqrt{x^2 + z^2}$. The energy $E = T + U$ is minimal when the pendulum is stationary in the downward position, $x = 0$, $z = -l$,

$$E_{\min} = -mgl + \frac{\kappa}{2}(l - l_0)^2 = -mg\left(l - \frac{mg}{2\kappa}\right). \tag{3.61}$$

There is no upper limit to the energy, which may increase to infinity. It is convenient [82] to introduce the dimensionless energy ratio R,

$$R = -\frac{E}{E_{\min}} \tag{3.62}$$

which varies from -1 to infinity.

When the bob starts to oscillate near its downward position, it is convenient to change its coordinates from $x = 0$, $z = -l$ to X and $z = -l - Z = -l_0 - g/\omega_s^2 - Z$. Substituting these changes into (3.60), one obtains for the Lagrangian of the oscillating system,

$$L = \frac{m}{2}\left[\left(\frac{dX}{dt}\right)^2 + \left(\frac{dZ}{dt}\right)^2\right] - mg\left(Z + l_0 + \frac{g}{\omega_s^2}\right) - \frac{\kappa}{2}(R_1 - l_0)^2 \tag{3.63}$$

with $R_1 = \sqrt{X^2 + (Z + l_0 + g/\omega_s^2)^2}$. The Lagrangian equations of motion are

$$\frac{d^2 X}{dt^2} = -\omega_s^2 \left(1 - \frac{l_0}{R_1}\right) X, \qquad (3.64)$$

$$\frac{d^2 Z}{dt^2} = -\omega_s^2 \left(1 - \frac{l_0}{R_1}\right)\left(Z + l_0 + \frac{g}{\omega_s^2}\right) - g. \qquad (3.65)$$

The quantity $1 - l_0/R_1$ can be rewritten as

$$1 - \frac{l_0}{R_1} = 1 - \frac{l_0}{(l_0 + g/\omega_s^2 + Z)}\left[1 + \frac{X^2}{(l_0 + g/\omega_s^2 + Z)^2}\right]^{-1/2}. \qquad (3.66)$$

Substituting (3.66) into (3.64) and (3.65) yields

$$\frac{d^2 X}{dt^2} = -\omega_s^2 X \left\{1 - \frac{1}{(1+\omega_0^2/\omega_s^2+Z/l_0)}\left[1 + \frac{X^2}{l_0^2\left(1+\omega_0^2/\omega_s^2+Z/l_0\right)^2}\right]^{-1/2}\right\}, \qquad (3.67)$$

$$\frac{d^2 Z}{dt^2} = -g - \omega_s^2\left(Z + l_0 + g/\omega_s^2\right) \qquad (3.68)$$

$$\times \left\{1 - \frac{1}{(1 + \omega_0^2/\omega_s^2 + Z/l_0)}\left[1 + \frac{X^2}{l_0^2\left(1 + \omega_0^2/\omega_s^2 + Z/l_0\right)^2}\right]^{-1/2}\right\}.$$

Thus far, all calculations have been exact. For small oscillations, $X < l_0$, one can expand the square root of (3.67) in a power series in X,

$$\frac{d^2 X}{dt^2} = -\omega_s^2 X \left[\frac{\omega_0^2/\omega_s^2 + Z/l_0}{1 + \omega_0^2/\omega_s^2 + Z/l_0} - \frac{\omega_s^6 X^2}{2l_0^2\left(\omega_s^2 + \omega_0^2 + \omega_s^2 Z/l_0\right)^3}\right]. \qquad (3.69)$$

If Z is also smaller than l_0, $(X, Z < l_0)$, then to lowest order in the small parameters X/l_0 and Z/l_0,

$$\frac{d^2 X}{dt^2} + \left[\frac{\omega_0^2\omega_s^2}{\omega_s^2 + \omega_0^2} + \frac{\omega_s^2}{l_0}\left(\frac{\omega_s^2}{\omega_s^2 + \omega_0^2}\right)^2 Z\right] X = 0. \qquad (3.70)$$

Performing the analogous calculations for Eq. (3.68), and inserting $g = \omega_0^2 l_0$, yields

$$\frac{d^2 Z}{dt^2} + \omega_s^2 Z = -\frac{\omega_s^6 X^2}{2l_0\left(\omega_s^2 + \omega_0^2\right)^2}. \qquad (3.71)$$

Without nonlinear terms, Eqs. (3.70) and (3.71) describe the "spring" and "oscillatory" modes of a spring pendulum, whereas the nonlinear term describes the simplest form of interaction between them.

Note that we have defined the oscillator frequency ω_0^2 relative to the constant parameter l_0, $\omega_0^2 = g/l_0$, in contrast to the frequency $\hat{\omega}_0^2$, which is defined relative to the variable length l, $\hat{\omega}_0^2 = g/l$, in order to keep the frequencies ω_0 and ω_s independent.

Using Eq. (3.56), one gets $\omega_s^2 - \hat{\omega}_0^2 = \omega_s^2 - g/\left(l_0 + g/\omega_s^2\right) = \omega_s^4/\left(\omega_s^2 + \omega_0^2\right)$. Then, one can rewrite Eqs. (3.70) and (3.71) in the following form,

$$\frac{d^2 X}{dt^2} + \hat{\omega}_0^2 X = \frac{\hat{\omega}_0^2 - \omega_s^2}{l} XZ, \tag{3.72}$$

$$\frac{d^2 Z}{dt^2} + \omega_s^2 Z = \frac{\hat{\omega}_0^2 - \omega_s^2}{2l} X^2, \tag{3.73}$$

which agrees with the equations of motion for the spring oscillator obtained by a slightly different method [83]. Note that Eqs. (3.72)-(3.73) represent two coupled oscillators X and Z with the simplest nontrivial coupling between their Hamiltonians of the form $X^2 Z$, which arises in other fields as well.

As it will be shown in Sec. 5.5, for the special relation $\omega_s = 2\hat{\omega}_0$, the spring and the pendulum are in autoparametric resonance while the energy transfers back and force from the string mode to the oscillating mode. For this case, Eqs. (3.72) and (3.73) take the following form,

$$\frac{d^2 X}{dt^2} + \hat{\omega}_0^2 X = -\left(3\hat{\omega}_0^2/l\right) XZ; \qquad \frac{d^2 Z}{dt^2} + 4\hat{\omega}_0^2 Z = -\left(3\hat{\omega}_0^2/2l\right) X^2. \tag{3.74}$$

Since these equations were obtained under the constraint $X, Z < l_0$, one can solve them using perturbation theory, assuming that the solution of the homogeneous equation oscillates with slowly-varying amplitude and phase [84],

$$X = A\left(t\right) \cos\left[\hat{\omega}_0 t + \psi\left(t\right)\right], \qquad Z = B\left(t\right) \cos\left[2\hat{\omega}_0 t + \chi\left(t\right)\right]. \tag{3.75}$$

Substituting (3.75) into (3.74) and neglecting the small nonresonant driving force, one obtains, to first-order in small derivatives of A, B, ψ and χ, the terms linear in cos and sin. Equating the coefficients separately to zero gives

$$\frac{dA}{dt} = \frac{3}{4l}\hat{\omega}_0 AB \sin\left(2\psi - \chi\right); \qquad \frac{dB}{dt} = -\frac{3}{16l}\hat{\omega}_0 A^2 \sin\left(2\psi - \chi\right); \quad (3.76)$$

$$\frac{d\psi}{dt} = \frac{3}{4l}B\cos\left(2\psi - \chi\right); \qquad \frac{d\chi}{dt} = \frac{3A^2}{16lB}\hat{\omega}_0 \cos\left(2\psi - \chi\right). \qquad (3.77)$$

Eliminating the sin factor from Eqs. (3.76) yields

$$\frac{d}{dt}\left(A^2 + 4B^2\right) = 0 \qquad (3.78)$$

which expresses conservation of energy,

$$A^2 + 4B^2 = M_0^2. \qquad (3.79)$$

From Eqs. (3.77), one obtains the second constant of motion

$$A^2 B \cos\left(2\psi - \chi\right) = N_0. \qquad (3.80)$$

The time dependence of the amplitude A can be found [84] from Eqs. (3.76), (3.79) and (3.80),

$$\frac{1}{2}\left(\frac{d\alpha}{d\tau}\right)^2 + V\left(\alpha\right) = E \qquad (3.81)$$

where E is a constant, and

$$\alpha = \frac{A^2}{M_0^2}; \qquad \tau = \frac{3\hat{\omega}_0 M_0}{4\sqrt{2l}}t; \qquad V\left(\alpha\right) = -\alpha^2 + \alpha^3. \qquad (3.82)$$

According to (3.82) and (3.79), $0 < \alpha < 1$. The two turning points α_0 and α_1 of the function $\alpha = \alpha\left(t\right)$ are defined by the equation $V\left(\alpha\right) = \mid E \mid$, so that $0 < \alpha_0 < \alpha_1 < 1$. Thus, the amplitude A of the oscillatory mode increases from $M_0\sqrt{\alpha_o}$ to $M_0\sqrt{\alpha_1}$, and then decreases back to $M_0\sqrt{\alpha_o}$, and goes back and forth, while at the same time, the amplitude B of the spring mode decreases and increases according to Eq. (3.79).

The interesting analysis of the sequence of order-chaos-order transitions in the spring pendulum was performed [82] by the use of the two dimensionless control parameters, R and μ. The first parameter R was defined in

(3.62), whereas the second parameter μ is defined as

$$\mu = 1 + \frac{\omega_s^2}{\hat{\omega}_0^2}. \tag{3.83}$$

This parameter can take any value between 1 and infinity. The special value is $\mu = 5$, for which $\omega_s = 2\hat{\omega}_0$. As will be shown in Sec. 5.5, for $\omega_s = 2\hat{\omega}_0$, the spring and pendulum are in autoparametric resonance while the energy transfers back and force from the string mode to the oscillating mode. We consider the behavior of the system at the limiting values of R and μ.

Limit value $\mu \to 1$ can be approached when the spring is weak (small κ) or the bob is heavy (large m). One can then rewrite the equations of motion (3.67) and (3.68),

$$\frac{d^2X}{dt^2} + \omega_s^2 X = (\mu - 1)\, gX \left[X^2 + \left(Z + l_0 + g/\omega_s^2 \right)^2 \right]^{-1/2},$$

$$\frac{d^2Z}{dt^2} + \omega_s^2 Z + g = (\mu - 1)\, gY \left[X^2 + \left(Z + l_0 + g/\omega_s^2 \right)^2 \right]^{-1/2}. \tag{3.84}$$

For very small $\mu - 1$, these equations describe two weakly coupled harmonic oscillators which, in the limit $\mu = 1$, have two smooth elliptic orbits. For $\mu = 1$, $\hat{\omega}_0 = \omega_s$ which explains the appearance of this characteristic frequency in Eq. (3.84).

Limit value $\mu \to \infty$ or $\kappa \to \infty$ means the transition to a rigid rod, i.e., the spring pendulum becomes a simple pendulum with periodic solutions, as considered in Sec. 1.1.

Limit value $R \to (-1)$ corresponds to small oscillations (librations) near the downward position. In this case, the equations of motion take the form (3.72)-(3.73), which define two coupled oscillators. However, for small deviations from the downward position, the coupling is very small which leads to simple solutions for harmonic oscillators.

Limit value $R \to \infty$ or $E \to \infty$ results in two equivalent rotations, one clockwise and the other counterclockwise, for which the energies of two modes, elastic and oscillatory, are the same order of magnitude [82].

In spite of the fact that the original equations contain chaotic solutions, which we will consider in Part 4, these solutions may appear only in the intermediate regions of the control parameters (Fig. 3.6). Therefore, changing each of these parameters, leads to transitions from regular to chaotic and then back to regular solutions.

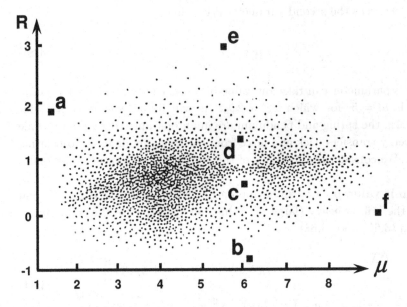

Fig. 3.6   The $(\mu - R)$ plane. The points a, b, e, f correspond to regular solutions while the shadow region (points c, d) describes chaotic solutions.

3.8   Resonance-type phenomena

3.8.1   *Stochastic resonance (SR)*

The standard definition of SR is the non-monotonic dependence of the signal-to-noise ratio on the strength of the noise [85]. We favor a wider definition of the SR as the non-monotonic behavior of an output signal, or some function of it, as a function of some characteristic of noise or of the periodic signal.

At first glance, it appears that all three ingredients, nonlinearity, periodic force and random force, are necessary for the appearance of SR. However, it has become clear that SR is generated not only in a typical two-well system, but also in a periodic structure [86]. Moreover, SR occurs even when each of these ingredients is absent. Indeed, SR exists in linear systems when the additive noise is replaced by nonwhite multiplicative noise [87]. Deterministic chaos may induce the onset of SR instead of a random signal [88]. Finally, the periodic signal may be replaced by a constant force in underdamped systems [89]. The underdamped pendulum, described by Eq. (3.12), was chosen as an example of an underdamped system [89]. SR manifests itself in the non-monotonic dependence of the average velocity $\langle d\phi/dt \rangle$ (or the mobility $\mu = a \langle d\phi/dt \rangle$) on the noise intensity D (see also [14]).

The mobility is plotted in Fig. 3.7. One sees from this graph [14] that μ has a SR-type peak close to the critical value $a_{cr} = 3.36b$. Note that in the absence of a periodic force, SR exists only in the limit of low damping. Moreover, the effect disappears in the overdamped case when one removes the second derivative in Eq. (3.12).

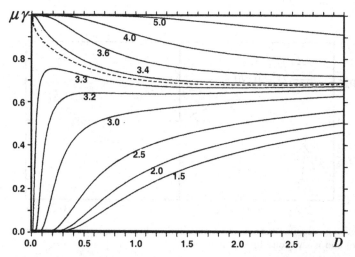

Fig. 3.7 Mobility μ times damping constant γ as a function of noise strength D, for various bias forces $\bar{a} = a/b$.

In the more conventional case of a pendulum subject to both constant and periodic torques, the average velocity has both a dc component $\langle d\phi/dt \rangle_{dc}$ described above, and an ac component $\langle d\phi/dt \rangle_{ac} = v_{ac} \sin(\omega t + \varphi)$, with $v_{ac} = fd(a\mu)/da$ and φ is the phase shift. Under the conditions $b \ll a$, and ω smaller than all characteristic times in the problem (adiabatic approximation), the maximum of v_{ac} is proportional to the amplification factor $(1 - a/a_{cr})^{-3/2}$.

In all the examples considered above, the pendulum was subject to a constant torque. There is no SR-type behavior without a constant torque [90].

3.8.2   *Absolute negative mobility (ANM)*

ANM means that a system subject to a bias force reacts in the direction
opposite to that of the acting force. Such behavior is forbidden for systems
which are in thermal equilibrium, but has been experimentally observed in
nonequilibrium systems such as multiple quantum-well structures [91] (see
also [92]). The theoretical analysis performed recently [72, 93] shows that
ANM also appears in a damped pendulum subject to constant, periodic
and random torques. Numerical simulation of the appropriate equation
has been made for the asymptotic mean velocity $\langle\langle v \rangle\rangle$ averaged over time
and thermal fluctuations.

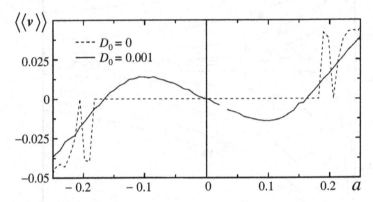

Fig. 3.8   Average angular velocity as a function of bias torque for the deterministic
(dashed line) and noisy (solid line) dynamics.

The results of these simulations are shown in Fig. 3.8 for the following
values of the dimensionless parameters: $\gamma = 0.9$, $f = 4.2$, $\omega = 4.9$. As one
can see, in the presence of noise, ANM occurs for the bias torque a in the
interval $(-0.17, 0.17)$. The need for both periodic torque and inertial effects,
expressed by the second derivative in Eq. (3.12), underlines ANM [72].
The physical reasons for this are clearly explained in [72, 93]. In the next
section, we consider the ratchet phenomenon which is closely connected
with ANM.

3.8.3   *Ratchets*

The term "ratchet" denotes a periodic potential which is anisotropic.
Ratchets produce a net flux without a driving force (bias force or exter-

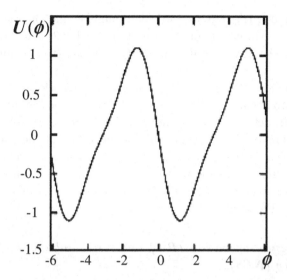

Fig. 3.9 Ratchet potential $U(\phi) = -\sin\phi - \frac{1}{4}\sin(2\phi)$.

nal gradient). Detailed information above ratchets can be found in a recent review article [45].

The equation of motion of a particle moving in a ratchet potential is given by

$$m\frac{d^2x}{dt^2} + \gamma\frac{dx}{dt} = -\frac{dV(x)}{dx} + f\sin(\omega t) + \xi(t) \tag{3.85}$$

where the anisotropic potential can be taken in the form shown in Fig. 3.9 [94],

$$V(x) = -\sin x - \frac{1}{4}\sin(2x). \tag{3.86}$$

In the overdamped case ($m = 0$) and in the absence of noise ($\xi = 0$), the solutions of Eq. (3.85) are either locked or running. In the latter case, the asymptotic value of the average velocity has the form

$$v_{qr} = \frac{x(t+rT) - x(t)}{rT} = \frac{q}{r}\omega \tag{3.87}$$

for integer r and q. For overdamped motion, $r, q > 0$, the flux is directed down the ratchet (positive current). The ratchet effect in a noiseless ratchet ($\xi = 0$ in (3.85)) is obtained only in the underdamped regime ($m \neq 0$ in (3.85)). In the latter case, the locked trajectories correspond to $r = q = 0$, whereas the running trajectories, both deterministic and chaotic, can occur, due to inertia, in either direction with $q = \pm1, \pm2, \ldots$ in Eq. (3.87). The direction of the flux depends sensitively on the control parameters in Eq. (3.85).

Strictly speaking, the ratchet effect means the transformation of the fluctuation environment into deterministic directional motion. In the case considered above, deterministically induced chaos mimics the role of noise. This result is supported by analysis [95] of the values of the control parameters for which current reversal occurs. The origin of current reversal is a bifurcation from a chaotic to a periodic regime. The same analysis has recently been extended [96] to the case in which an additional constant force is added to Eq. (3.85).

3.8.4 *Resonance activation (RA) and noise enhanced stability (NES)*

Both RA and NES are counterintuitive phenomena. The NES effect implies that noise can modify the stability of a system placed at the metastable state if its lifetime is greater than the deterministic lifetime [97]. The constructive role of noise also manifests itself in RA, with thermal noise assisting the crossing of temporally modulated potential barriers [44]. Previous theoretical and experimental studies were restricted to overdamped systems. Only recently were experiments performed [98] which show the existence of RA and NES in underdamped real systems (Josephson junctions). Numerical simulation yields good agreement with the experimental results [98].

Chapter 4

Deterministic Chaos

4.1 General concepts

One of the great achievements of twentieth-century physics, along with quantum mechanics and relativity, is the new relationship between order and disorder (chaos). Noise, which had always played a destructive role, becomes constructive. One can illustrate this phenomenon by the two well-known statements. First, "a butterfly that flaps its wings in China causes rain in Texas". Second, if you cannot hear your friend's voice in the telephone booth, wait for a passing police car or an ambulance, and their noise will help you in your telephone conversation. The meaning of these statements is that order and noise (disorder) are not contradictory but complimentary [99]. Such phenomena as "deterministic chaos" and "stochastic resonance", whose names are half-deterministic and half-stochastic, express the deep connection between these two seemingly opposite phenomena. Deterministic systems, described by differential equations, are fully defined by the initial conditions and, therefore, in principle remain fully predictable, no matter how many particles they contain and how strong is the interaction between these particles. However, an important property of nonlinear equations is the exponential increase in time of their solutions when one makes even the smallest change in the initial conditions. "Deterministic chaos" appears without any random force in the equations. Such a situation is very common since, in principle, an infinite number of digits is required to specify the initial conditions precisely, an accuracy that is obviously unattainable in a real experiment.

In order to exhibit deterministic chaos, the differential equations have to be nonlinear and contain at least three variables. This points to the important difference between underdamped and overdamped equations of motion of a pendulum. The underdamped equation, subject to an external periodic force,

$$\frac{d^2\phi}{dt^2} + \gamma\frac{d\phi}{dt} + \sin\phi = f\sin\left(\omega t\right),\qquad(4.1)$$

can be rewritten as a system of three first-order differential equations

$$\frac{d\chi}{dt} + \gamma\chi + \sin\phi = f\sin\left(\theta\right);\qquad(4.2)$$

$$\frac{d\phi}{dt} = \chi;\qquad\frac{d\theta}{dt} = \omega\qquad(4.3)$$

and, therefore, the underdamped equation exhibits deterministic chaos for some values of the parameters. On the other hand, the overdamped equation, in which $d^2\phi/dt^2 = 0$, has only two variables, and therefore it does not exhibit chaos. In this chapter we consider only the underdamped pendulum starting from Eq. (4.1).

There are many books describing deterministic chaos including [12], which considers chaos for the damped driven pendulum. We shall here present only those basic concepts which are needed in the following discussion.

4.1.1   *Poincare sections and strange attractors*

The hallmark of deterministic chaos is the sensitive dependence on initial conditions, which means that two trajectories having very similar initial conditions will diverge exponentially in time. The Poincare section is obtained by removing one space dimension. In our case of three variables, the Poincare section is a plane. The Poincare plane has to be chosen in such a way that trajectories will intersect it several times. If the motion is periodic (non-chaotic), the trajectory will cross the Poincare plane again and again at the same point. The number of points appearing in the Poincare plane defines the number of periods corresponding to a given trajectory. However, chaotic motion means that despite the fact that the motion is deterministic (i.e., for given initial conditions, the equations will exactly describe the trajectory), it never repeats itself. There will be a dense set of points on the Poincare section filling a certain area of this plane. The locus of these points is called a strange attractor, and all initial points that eventually bring the system to these attractors are called the basin of the attractor. In this way, one reduces the continuous trajectories in the phase space to a discrete mapping on the Poincare plane.

4.1.2 *Lyapunov exponent*

The Lyapunov exponent λ shows the extent to which the two nearby trajectories, separated initially by distance d, become separated in time by a large distance D,

$$D = d \exp(\lambda t). \tag{4.4}$$

In a chaotic regime, λ is positive, whereas in the regular regime, λ is negative or zero, implying that the initial separation either diminishes or remain constant. For a system of first-order differential equations, there are several Lyapunov indices, and we are interested in the largest of these.

4.1.3 *Correlation function*

Analysis of the autocorrelation function of a trajectory shows the difference between regular and chaotic regimes. In the latter case, the system loses information about previous states and the autocorrelation function tends to zero with time. However, since chaotic trajectories densely fill phase space, there is some short time during which the trajectory approaches the initial position, and the autocorrelation function may grow again. In contrast to these results, for the regular trajectory, the autocorrelation function oscillates near some average value, increasing and decreasing when the system moves away or approaches its initial position, respectively.

4.1.4 *Spectral analysis*

The squared modulus of the Fourier transform of the variable $g(t)$ is called the power spectrum of $g(t)$. Periodic motion is described by the power spectrum with a finite number of frequencies, whereas chaotic motion has a broad power spectrum. In Figs. 4.1 and 4.2, we show [100] the efficiency of the four methods of detecting chaos described above, using Eq. (4.1) as an example. A comparison between Figs. 4.1 and 4.2 shows that although some methods (power spectrum) are ambiguous, one can see a clear distinction between regular (Fig. 4.1) and chaotic (Fig. 4.2) behavior.

We now describe the two paths of the transition to chaos: period doubling and intermittency.

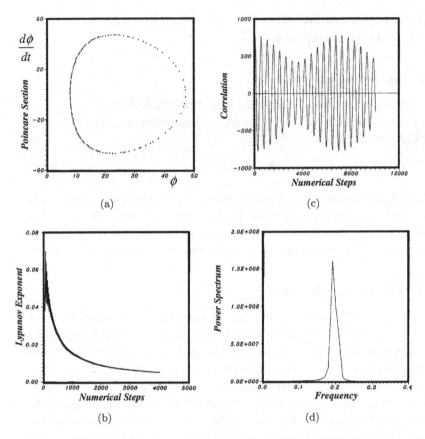

Fig. 4.1   Plot of several indicators for a regular trajectory: (a) Poincare section,
(b) Maximum Lyapunov exponent, (c) Correlations function, (d) Power spectrum.

4.1.5   *Period doubling and intermittency*

Starting from the periodic regime, the change of some parameters in the dynamic equations may move the system into the chaotic regime [101], [102]. The most common ways of entering the chaotic regime are through period doubling and intermittency. The former means that the chaos is preceded by successive period-doubling bifurcations with a universal rate in different systems. In contrast to period doubling, for intermittency the periodic solution remains stable up to the onset of chaos which manifests itself in short escapes from the periodic solution which take place at irregular intervals [103].

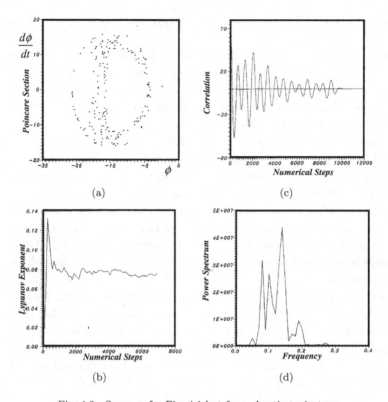

(a) (c)

(b) (d)

Fig. 4.2 Same as for Fig. 4.1 but for a chaotic trajectory.

4.2 Transition to chaos

4.2.1 *Damped, periodically driven pendulum*

The noise phenomena of Eq. (4.1), written in dimensionless time $\tau = t/\alpha$,

$$\frac{d^2\phi}{d\tau^2} + \alpha\gamma\frac{d\phi}{d\tau} + \alpha^2 \sin\phi = f\alpha^2 \sin(\alpha\omega\tau), \qquad (4.5)$$

have been analyzed in a pioneering work [104]. Equation (4.5) was solved for $\alpha\gamma = 0.5$ and $\alpha^2 = 6.4$ for different amplitudes $f\alpha^2$ and frequencies $\alpha\omega$, which is the ratio of the external frequency to the eigenfrequency of the system. For $\alpha\omega << 1$ and $\alpha\omega >> 1$, the solutions of Eq. (4.5) are periodic [104]. However, the solutions can be quite complicated with subharmonics, harmonics, hysteresis loops, etc. For different regions of the $f\alpha^2 - \alpha\omega$ plane, one obtains locked or running solutions. However, there exist a region of this plane, located near $f\alpha^2 = 3.8$ and $\alpha\omega = 0.64$, where the solutions exhibit the set of period-doubling cascading bifurcations into a chaotic state with strange attractors.

In addition to the period-doubling scenario, another path to chaos, intermittency, was found [105] by plotting ϕ as function of t for parameters $\alpha = 1$, $\gamma = 0.5$, $\omega = 0.47$, and different values of f. The characteristic parameters of intermittency found by numerical simulation and analog studies agree with theoretical predictions [103].

The intermittency scenario of the transition to chaos in Eq. (4.5) was also found [106] for $\alpha = 1$, $\gamma = 0.5$, $\omega = 2/3$ and different amplitudes f of an external field. For $f \leq 1.5$, there exist two separate stable periodic running solutions, clockwise and counterclockwise. As f increases, these two modes remain separate while becoming chaotic. At the critical value $f = 1.4945$, intermittent switching between these two modes occurs, producing a large amount of noise at frequencies smaller than the driving frequency ω. These results have been proved by analyzing the $(\phi, d\phi/dt)$ phase portrait and the Poincare sections.

Parallel numerical calculations of the onset of chaos and the Fourier spectrum $(d\phi/dt)_\omega$ of the angular velocity show [107] that at the same values of the parameters which define the onset of chaos, there also appears a large gain in the parametric amplifier manifested as a maximum of $(d\phi/dt)_\omega$ (voltage in Josephson junctions). Note that for different values of the parameters, this large gain appears without the onset of chaos.

The comprehensive analysis [108] of Eq. (4.5) revealed many other peculiarities in the periodic-chaotic transitions described by Eq. (4.5). Calculations of $\phi(t)$ as a function of the amplitude and frequency of the periodic torque at constant damping, show quite different behavior for low and high dissipation. At low dissipation, there exist broad bands of chaotic solutions where the transition to chaos might proceed through both double frequency bifurcation and intermittency, depending on the symmetry of the equation with respect to the simultaneous phase inversion and the shift of the phase of the driving force by an odd multiple of π. No chaotic states have been found for high dissipation.

The extensive analysis [109] of equation (4.5), written in dimensionless form as

$$\frac{d^2\phi}{d\tau^2} + \gamma\frac{d\phi}{d\tau} + \sin\phi = f\sin\left(\frac{\omega}{\omega_0}\tau\right), \qquad (4.6)$$

where $\tau = \omega_0 t$, shows that the ratio of the eigenfrequency of a pendulum $\omega_0 = \sqrt{g/l}$ and the external frequency ω plays a crucial role for the transition to chaos. The solutions of Eq. (4.6) can be divided in different groups. Analogously to the solutions (2.4) for the overdamped pendulum, Eq. (2.1) without noise, there are running and locked solutions of Eq. (4.6). In addition, there are symmetric solutions to the downward positions and symmetry-breaking solutions which oscillate with a larger amplitude to one side than to the other. Three different paths to chaos have been found [109]: (1) Period-doubling cascade preceded by the appearance of symmetry-breaking solutions. This cascade is produced in both running and locked solutions. In the latter case, the frequency ω_0 is locked to ω. (2) The loss of phase locking and random transitions between two locked states. (3) The intermittency form of the transition to chaos occurs for $\omega \ll \omega_0$, but for large amplitude f of an external field (larger than the amplitude defining the transition from locked to running solutions). The trajectories are then a combination of clockwise and couterclockwise rotations with damped oscillations in-between.

The following interesting question has been asked [110]. Since both deterministic chaos and a random force are able to produce chaotic behavior, what would be the result of the joint action of both these factors? To answer this question, Eq. (4.5) has been supplemented by additive white noise $\eta(t)$ of strength D which transforms Eq. (4.5) into the following equation

$$\frac{d^2\phi}{dt^2} + \gamma\frac{d\phi}{dt} + \omega_0^2 \sin\phi = f\cos\omega t + \eta(t). \tag{4.7}$$

To transform Eq. (4.7) in a form for which the Fokker-Planck equation can easily be written, the following change of variables is introduced,

$$x_1 = \phi; \qquad x_2 = \frac{d\phi}{dt}; \qquad x_3 = f\cos\omega t; \qquad x_4 = \frac{dx_3}{dt}. \tag{4.8}$$

Equation (4.7) then takes the following form

$$dx_1/dt = x_2; \qquad dx_2/dt = -\gamma x_2 - \omega_0^2 \sin\phi + x_3 + \eta(t);$$
$$dx_3/dt = x_4; \qquad dx_4/dt = -\omega^2 x_3. \tag{4.9}$$

In this case, the Focker-Planck equation for the distribution function $P(x_1, x_2, x_3, x_4)$ (with the initial conditions $x_{1,0}$ and $x_{2,0}$),has the following form,

$$\frac{\partial P}{\partial t} = -\frac{\partial}{\partial x_1} [x_2 P] - \frac{\partial}{\partial x_2} \left[\left(-\gamma x_2 - \omega_0^2 \sin x_1 + x_3 \right) P \right]$$
$$- \frac{\partial}{\partial x_3} [x_4 P] - \frac{\partial}{\partial x_4} \left[-\omega^2 x_3 P \right] + \frac{1}{2} D \frac{\partial^2 P}{\partial x^2}. \tag{4.10}$$

Equation (4.10) was solved numerically for $\gamma = 1.0$, $\omega_0^2 = 4.0$, $\omega = 0.25$, $D = 0.5$ for two sets of initial conditions: $x_{1,0} = 0.0$, $x_{2,0} = 0.0$ and $x_{1,0} = 0.0$, $x_{2,0} = 0.1$. For these values, the stationary state distribution function P shows chaotic behavior [110]. Note that for these parameters, Eq. (4.7) without noise also shows chaotic behavior [111].

Quite different results were obtained [112], [113] for the case in which Eq. (4.5) is supplemented by multiplicative white noise in the form of the fluctuating frequency Ω of an external field. The dimensionless form of this equation is

$$\frac{d^2\phi}{d\tau^2} + \gamma \frac{d\phi}{d\tau} + [(1 + f \cos(\Omega t))] \sin \phi = 0. \tag{4.11}$$

Upon adding noise to the frequency, the initial chaotic motion becomes regular and terminates at one of the fixed points. Calculations have been performed [112] for high-frequency stochastic oscillations for the following parameters: $f = 2$ and $f = 8$, $\gamma = 1$, $\omega_0 \equiv \omega/\Omega = 0.5\pi$, for two sets of initial conditions: $\phi(0) = 0$, $d\phi/dt(0) = 1$ and $\phi(0) = 2$, $d\phi/dt(0) = 0$. In the former case, one obtains nonregular librations without noise, whereas adding noise terminates the motion at the fixed point $\phi_f = 0$. In the latter case, nonregular librations and rotations are terminated at the fixed point $\phi_f = -2\pi$. If a system originally had a stable limit cycle instead of fixed points, then the addition of noise will terminate the chaotic behavior into a regular limit cycle. Analogous numerical calculations (with $\sin \phi$ replaced by $b\phi + c\phi^3$) have been performed [113] for low-frequency stochastic oscillations for the parameters: $f = 0.94$, $\gamma = 0.15$, $\omega_0 \equiv \omega/\Omega = 0.5\pi$ and initial conditions $\phi(0) = 1$, $(d\phi/dt)_0 = 1$. The results were similar to those obtained [112] without noise: the motion is chaotic with randomly alternating librations and rotations, whereas adding sufficient random noise stabilizes the system by eliminating chaos. The latter result has been confirmed [113] by the calculation of the Lyapunov indices.

It is remarkable that in contrast to additive noise, multiplicative noise is able to convert the chaotic trajectories, induced by the deterministic chaos, into regular trajectories.

4.2.2 *Driven pendulum subject to a periodic and constant torque*

The addition of a constant torque a to Eq. (4.1) yields

$$\frac{d^2\phi}{dt^2} + \gamma\frac{d\phi}{dt} + \sin\phi = a + f\sin(\omega t), \tag{4.12}$$

which permits the comparison of the onset of chaos with the analysis of the Shapiro steps in the $(\langle d\phi/dt\rangle, a)$ plane performed in Sec. 2.3.1. The road to chaos was found [114] to be different for the regions with increasing values of $\langle d\phi/dt\rangle$ ("running" solutions) and for those with Shapiro steps, which correspond to constant values of $\langle d\phi/dt\rangle$ ("locked" solutions). In the latter case, the system exhibits period-doubling bifurcations, whereas in the former case, the transition to chaos goes through the intermittency scenario.

The voltage-current graph ($\langle d\phi/dt\rangle$ versus a) for different damping constants has been analyzed [109]. For $f = 0$ and $a \gg 1$, the angular velocity $\langle d\phi/dt\rangle$ is very high, and one can neglect the nonlinear sinusoidal term in (4.12). For this case, the pendulum reaches the asymptotic value $\langle d\phi/dt\rangle = a/\gamma$ in a time of order γ^{-1}. For $\gamma \leq 1$, the voltage-current graph exhibits hysteresis which allows a simple physical explanation [109]. Indeed, when the bias a increases from zero to unity, due to the inertia term, the pendulum starts to rotate, approaching the limiting value $\langle d\phi/dt\rangle = 1/\gamma$. On the other hand, upon decreasing a, the inertia causes the pendulum to continue rotating even for $a < 1$, until reaching the critical value a_{cr} where the angular velocity vanishes, and the pendulum relaxes toward the equilibrium position. For $\gamma \to 0$, $a_{cr} \approx \sqrt{2\gamma}$. No fundamental difference occurs when $f \neq 0$, i.e., when both constant and periodic torques are present.

A comprehensive analysis has been performed [115] of the dimensionless Eq. (4.12), written in the slightly different form,

$$\beta\frac{d^2\phi}{dt^2} + \frac{d\phi}{dt} + \sin\phi = a + f\sin(\Omega t). \tag{4.13}$$

The simplest solutions of Eq. (4.13) are the steady-state periodic solutions $\phi(t)$, with driving frequency Ω which obeys the relation

$$\phi\left(t + \frac{2\pi m}{\Omega}\right) = \phi(t) + 2\pi l. \tag{4.14}$$

The average angular frequency is $\langle d\phi/dt\rangle = l\Omega/m$, with integer l and m. Solutions with $m = 1$ are harmonic, those with $1 < m < \infty$ are subharmonic, and those with $m = \infty$ are chaotic. These three regions can be distinguished [115] by the power spectrum of the angular velocity

$$S(\omega) = \frac{2}{T} \int_0^T \frac{d\phi}{dt} \exp(i\omega t)\, dt. \tag{4.15}$$

Chaotic solutions appear only in restricted regimes of the parameters β, a, f and Ω, in the $(d\phi/dt, \phi)$ phase diagram. One can formulate [115] some general criteria for the onset of chaos. For example, chaos does not occur for $\beta \ll 1$, which is close to an overdamped case, or for $\Omega \gg \beta^{-1/2}$, which means that the driving frequency is far from the eigenfrequency of the pendulum.

4.2.3 *Pendulum with vertically oscillating suspension point*

As we will see in Part 5, vertical oscillations of the suspension point is one way to transform the unstable upward position of a pendulum into a stable position. We return to this case in order to consider the transition to chaos. Equation (3.29), supplemented by damping, has the following form

$$\frac{d^2\phi}{dt^2} + \gamma\frac{d\phi}{dt} + [a + 2f\cos(\omega t)]\sin\phi = 0. \tag{4.16}$$

The first numerical analysis of Eq. (4.16) was performed in 1981 by McLaughlin [116] for $a = 1$, $\omega = 2$ and three values of γ: 0.2, 0.05, and 0. In order to obtain motion with finite amplitude, the control parameter f has to satisfy the condition $f \geq \gamma$ [117]. For $\gamma = 0.2$, simple vibration of period 2π is the only solution for $0.2 < f < 0.713$. At these values of f, two rotations of period π appear. Then, at $f = 0.7925$, these rotations go through a period-doubling bifurcation according to the Feigenbaum scenario [101], until finally, a pair of strange attractors appear. Up to $f = 1.045$, the sign of $(d\phi/dt)$ is negative (clockwise rotation), but for larger f, the angular velocity occasionally switches sign. The strange attractors exist up to $f = 1.58$. Beyond this point, there are two stable solutions of period π. Thus, increasing the control parameter f causes the system to go through the transformation order-chaos-order. Qualitatively similar behavior takes place for the damping constant $\gamma = 0.05$. No strange attractors exists for the Hamiltonian case $\gamma = 0$, and the double-frequency bifurcation leads to the destruction of stable zones.

The numerical solution of Eq. (4.16) has been carried out [118] for different sets of parameters. For $\gamma = 0.15$, $a = 1$ and $\omega = 1.56$, the results are similar to those obtained in [116], with fully chaotic motion for $f > 0.458$. The behavior of a system near the latter point depends on the initial conditions [118]. For $\alpha = 0.5$, $\gamma = 0.03$ and varying f, a double-frequency bifurcation begins [74] when $f = 0.55$, with the motion being fully chaotic for $f = 0.64018$.

A comprehensive numerical solution of Eq. (4.16) has been carried out [119] for both the amplitude and the frequency of the varying external field. For $\gamma = 0.2\pi$ in Eq. (4.16), upon increasing f, the system undergoes an infinite series of period-doubling transitions to chaos for all ω. Moreover, for each ω, there exists an infinity of alternating stable and unstable regimes of f, similar to Mathieu-type diagrams.

The interesting analysis of the concurrent action of deterministic chaos and noise, induced by adding a random force to Eq. (4.16), has been performed [120], [121]. Noise introduces hops between the periodic attractors, whereas in the noise-free system the pendulum remains fixed in one of these attractors [121]. On the other hand, external noise does not influence the chaotic transient, i.e., the characteristic time for the transition from chaotic to periodic trajectories with increasing control parameter [120].

Another way to introduce noise into Eq. (4.16) has been exploited [75], where a random phase φ was included in the equation by replacing $\cos(\omega t)$ by $\cos(\omega t + \varphi)$. For weak noise, transitions from oscillations to rotations and the opposite are possible. For moderate noise, there is a combination of the excitation amplitude and stochastic component. Finally, for strong noise, there is no rotational mode, and the transition to chaos occurs via intermittency.

4.2.4 *Pendulum with horizontally oscillating suspension point*

The equation of motion (3.47) for the case has the following form,

$$\frac{d^2\phi}{d\tau^2} + \left(\frac{k}{r}\right)\frac{d\phi}{d\tau} + \frac{1}{r^2}\sin\phi - f\sin\tau\,\cos\phi = 0 \qquad (4.17)$$

where $r = \omega/\omega_0$ is the ratio of the frequency of an external field ω to the pendulum frequency $\omega_0 = \sqrt{g/l}$, $F = f/l$ is the dimensionless amplitude of an external field, and $\tau = \omega t$.

Equation (4.17) is symmetric with respect to the following transformations

$$\phi \longrightarrow -\phi; \qquad \frac{d\phi}{d\tau} \to -\frac{d\phi}{d\tau}; \qquad \tau \to \tau + \frac{\pi}{2}. \qquad (4.18)$$

The trajectory in phase space $(\phi, d\phi/dt)$ that is invariant to these symmetry transformation is called a symmetric trajectory. Otherwise, it is called an asymmetric trajectory. Extensive numerical analysis of Eq. (4.17) has been performed [122] for different values of the amplitude f of an external field in the interval $f \in (0, 15)$, for $k = 0.1$ and $r = 0.8$. Five cascades of double-frequency bifurcations were found for specific values of f which satisfy Feigenbaum theory [101]. Moreover, at the limiting value $f = 1.4336$, the behavior of the system becomes chaotic. The appearance of frequency-doubling and the transition to chaos can be clearly seen in each cascade. For the first cascade, the transitions periodic orbit $\to$ frequency doubling $\to$ chaos can be clearly seen from the $(\phi, d\phi/dt)$ phase plane shown in Fig. 4.3.

Additional information has been obtained [122] by studying the Lyapunov exponents, the power spectrum and the evolution of strange attractors with the change of the control parameter f. The solutions of Eq. (4.17) for different frequencies of an external field at constant amplitude has been studied numerically [123]. As the frequency is decreased, for sufficiently large amplitude, the system progresses from symmetric trajectories to a symmetry-breaking period-doubling sequence of stable periodic oscillations, and finally to chaos.

The analysis of Eq. (4.17) for the special case of small amplitude and high frequency of the external field has been performed [124]. The latter conditions are met if one replaces the frequency ω of the external field by Ω such that $\Omega = \omega/\varepsilon$. The small parameter ε also determines the smallness of the amplitude, $f = \varepsilon l \beta$, and of the damping $k/l = 2\alpha\varepsilon$. In terms of these new parameters and the dimensionless time $\tau = \omega t$, Eq. (4.17) takes the following form

$$\frac{d^2\phi}{d\tau^2} + 2\alpha\varepsilon\frac{d\phi}{d\tau} - \beta\varepsilon \sin \tau \cos \phi + \varepsilon^2 \sin \phi = 0. \qquad (4.19)$$

According to the method of multiple scales [125], one seeks the solution of Eq. (4.19) in the form

$$\phi = \phi_0 + \varepsilon\phi_1 + \varepsilon^2\phi_2 + \cdots \qquad (4.20)$$

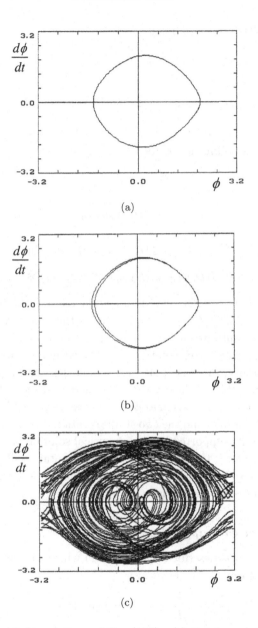

(a)

(b)

(c)

Fig. 4.3 Phase plot for solutions of Eq. (4.17): (a) Regular solution with period π for $f = 1.429$ with initial conditions $\phi_0 = 0.056841$ and $(d\phi/dt)_0 = -2.050299$, (b) Regular solution with period 2π for $f = 1.4359$ with initial conditions $\phi_0 = 0.140078$ and $(d\phi/dt)_0 = -2.033011$, (c) Chaotic trajectory for $f = 1.45$ with cycles of period 2π.

where ϕ_n is a function of $\varepsilon^n \tau_n$. Then,

$$\frac{d}{d\tau} = D_0 + \varepsilon D_1 + \varepsilon^2 D_2 + \cdots$$

$$\frac{d^2}{d\tau^2} = D_0^2 + 2\varepsilon D_0 D_1 + 2\varepsilon^2 D_0 D_2 + \varepsilon^2 D_1^2 \cdots$$

(4.21)

where $D_n \equiv d/d\tau_n$. Substituting Eqs. (4.20) and (4.21) into (4.19) yields three equations, containing the coefficients ε^0, ε^1 and ε^2, respectively

$$D_0^2 \phi_0 = 0,$$

(4.22)

$$D_0^2 \phi_1 = -2D_0 D_1 \phi_0 + \beta \sin \tau_0 \cos \phi_0 - 2\alpha D_0 \phi_0,$$

(4.23)

$$D_0^2 \phi_2 = -2D_0 D_1 \phi_1 - 2D_0 D_2 \phi_0 - D_1^2 \phi_0 - \sin \phi_0$$
$$- 2\alpha D_0 \phi_1 - 2\alpha D_1 \phi_0 - \beta \phi_1 \sin \tau_0 \cos \phi_0.$$

(4.24)

The solution of Eq. (4.22) that does not increase with time is $\phi_0 = Const$. Using Eqs. (4.23)-(4.24), and eliminating the secular (unbounded) terms yields three solutions for the stationary state, $\phi_0 = 0$, $\phi_0 = \pi$ and $\phi_0 = \cos^{-1}\left(2/\beta^2\right)$. The latter solution is obtained under the assumption that $\beta^2 \geq 2$. Stability analysis [124] shows that the fixed point solution $\phi_0 = \pi$ is always unstable, whereas for $\beta^2 < 2$ the only stable point is $\phi = 0$. When the amplitude of the external force f increases beyond $\beta^2 > 2$ the zero solution becomes unstable, and the two stable solutions are $\pm \cos^{-1}\left(2/\beta^2\right)$. This analytical result, supported by the numerical solution of Eq. (4.19), is quite surprising. Indeed, the pendulum with horizontal oscillations of the suspension point does not perform oscillations around the horizontal axis, but rather around the inclined axis!

4.2.5 *Pendulum with applied periodic force*

The forced pendulum is described by Eq. (3.21) which can be rewritten in the dimensionless form as

$$\frac{d^2\phi}{d\tau^2} + \gamma \frac{d\phi}{d\tau} + \sin \phi = f \sin (\Omega t)$$

(4.25)

with Ω being the ratio of the external frequency ω to the characteristic frequency $\sqrt{g/l}$ of the pendulum. Numerical simulation shows [126] that for some values of the parameters, the solutions of Eq. (4.25) are oscillations around the upward position. For $\gamma = 0.25$ and $f = 2.5$, there are stable inverted oscillations for $0.725 < \Omega < 0.792$. Symmetry-breaking, period-doubling and chaotic modulations occur for $\Omega = 0.794$, 0.796 and 0.800, respectively. Many subharmonics have been found for smaller values of Ω [127].

When the amplitude of the external field f becomes non-zero, the stable stationary point $\phi = d\phi/dt = 0$ is replaced by a stable symmetric periodic trajectory. Detailed analysis of the bifurcations of this trajectory has been performed [128], showing alternating stable and unstable behavior upon varying the control parameters f and ω, for a fixed damping constant $\gamma = \pi$.

Detailed numerical calculations show [129] a plethora of trajectories at periods equal to the forcing period or its integral multiples, stable running trajectories with mean angular velocity $p\omega/q$ with integers p and q, and period-doubling cascades leading to chaotic motion.

A convenient classification of solutions of Eq. (4.25) has been suggested [130] using the language of the Brownian particle described by Eq. (4.25). Divide the coordinate ϕ into two parts

$$\phi = 2\pi \left(n + \psi \right) \tag{4.26}$$

where n and ψ are the integer part and the non-integer part of ϕ, respectively. The potential $U(\phi) = -\cos\phi$ has periodicity 2π. Therefore, n indicates the location of the valley in which the particle is situated, and ψ represents the position of the particle within the valley. One can distinguish several different cases:

Variable ψ is periodic. (a) N-point intra-valley motion (fixed n); (b) N-point intra-valley motion (periodic n); (c) N-point drift motion (increasing or decreasing n).

Variable ψ is chaotic. (a) Intra-valley chaotic motion (fixed n); (b) Inter-valley motion without diffusion and drift (periodic n); (c) The same without diffusion but with drift (increasing or decreasing n); (d) The same with diffusion but without drift (chaotic n); (e) The same with diffusion and drift (chaotically increasing or decreasing n).

The interesting idea has been proposed [131] to use the locked chaotic solutions of Eq. (4.25) to generate white noise. Using heuristic arguments confirmed by numerical calculations, it was proved that this high-level noise is white over many decades in frequency. The output voltage $d\phi/dt$ is usually locked with the external signal, so that $\langle d\phi/dt \rangle = \Omega$. However, for appropriate values of the parameters γ, f and Ω, the power spectrum of $d\phi/dt$ includes a broadband component which is just the required white noise. Because of the exact periodicity and uniformity of the potential $U(x) = -\cos\phi$, the Brownian particle forgets its location and limits the time over which the correlation occurs. The absence of correlations implies white noise. Therefore, the diffusion is normal, in contrast to the anomalous diffusion considered in Sec. 3.3. For the numerical simulations of Eq. (4.25), the following values of the parameters were used [131]: $\gamma = 0.1, f = 1.6$, $\Omega = 0.8$. The analysis of 20 chaotic trajectories with different initial conditions clearly shows the diffusive character of the motion with a positive Lyapunov index $\lambda = 0.137$, the Gaussian distribution of ten thousand trajectories with standard deviations proportional to $\sqrt{t}$, and the power spectrum (4.15) shown in Fig. 4.4. Together with discrete components at the driving frequency and its harmonics, the latter clearly establishes that the white portion of the spectrum extends over four decades in frequency.

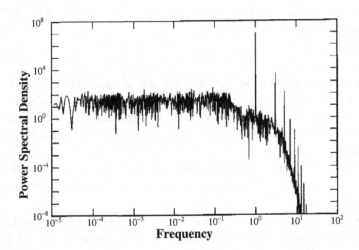

Fig. 4.4   Power spectral density of the average angular velocity as a function of frequency for the chaotic regime, computed over 2^{19} drive cycles. Parameters are $\gamma = 0.1$, $\Omega = 0.8$, $f = 1.6$.

4.2.6 *Spring pendulum*

The equations of motion of a spring pendulum, (3.64) and (3.65), were analyzed [82] for different values of the control parameters R and μ, defined in (3.62) and (3.83), respectively. As described in Sec. 3.7, chaotic states appear for intermediate values of R and μ, while the regular non-chaotic solutions occur for the limiting values of these parameters. These types of solutions are shown [82] in the $R - \mu$ plane in Fig. 3.6. The shading in the central region of this plane indicates the region of chaotic solutions. The points marked from a to f show the regular solutions in points a, b, e, f, whereas points c and d represent the chaotic solutions. Chaos is connected with the coupling of two degrees of freedom, so that the point $\mu = 5$, which corresponds to autoparametric resonance, is a natural source of chaos. The boundary between the locked and running solutions is another region where the chaotic solutions are clustered. For a rigid pendulum, the boundary is defined by the separatrix, whereas for a spring pendulum, these boundaries are described [82] by the following curves in the $R - \mu$ plane, below which there is no running solutions,

$$R = 1 - 2\left[\left(2\left(\mu - 1\right) + 1\right)\right]^{-1}, \qquad \text{for } \mu > 2,$$

$$R = \left(\mu - 1\right)\left[2 + \left(\mu - 1\right)^{-1}\right]^{-1}, \qquad \text{for } 1 < \mu < 2. \tag{4.27}$$

At $\mu = 2$, where $\omega_s = \omega_0$, these two curves coincide showing the borderline for running solutions.

4.3 Pendulum subject to two periodic fields

4.3.1 *Controlling chaos*

Chaotic behavior is very sensitive not only to small changes in initial conditions, but also to small additional forces acting on the pendulum. The latter feature permits controlling and even suppressing chaos (see the review [132] and the recent article [133] with many references therein). As an illustration, let us consider the possibility of eliminating chaos by applying an additional periodic field $A \sin\left(\beta \omega t\right)$ to the dynamic equation (4.12), [134]

$$\frac{d^2\phi}{dt^2} + \gamma \frac{d\phi}{dt} + \sin\phi = a + f \sin\left(\omega t\right) + A \sin\left(\beta \omega t\right). \tag{4.28}$$

For $A = 0$, this equation shows [135] chaotic behavior for the following values of the parameters: $\gamma = 0.7$, $f = 0.4$, $\alpha = 0.715$, and $\omega = 0.25$. The resulting dynamics as a function of the parameters A, β and a. has been obtained [134] by numerical calculation of the Lyapunov exponents for different values of these parameters. The influence of the frequency of an additional periodic force has been studied by changing the parameter β for $a = 0.905$ and $A = 0.0125$. There are regimes of β, say, $0.18 < \beta < 0.6$, where the trajectories are regular. This shows that in order to reduce or eliminate chaos, it is enough to apply to the system a weak time-dependent periodic perturbation of small amplitude $A = 0.0125$. These results have recently been refined [136]. In order to change the dynamic behavior, one has to supply energy in order to overcome the characteristic energy of the system. The latter is determined by the parameter a which defines the height of the barrier of the potential well. Therefore, a "weak" perturbation means that the perturbation is small for small values of a [134]. Careful study shows [136] that there are two critical (depending on f, ω and β) values of the amplitude of the external field, $A_{cr.1}$ and $A_{cr.2}$, such that for $A < A_{cr.1}$, the increase of A increases rather than decreases the chaos. Only for $A > A_{cr.2}$, does an increase of A eliminates chaos. The reaction of the system for $A_{cr.1} < A < A_{cr.2}$ is more complicated and depends on the values of the parameters.

4.3.2   *Erratic motion*

Chaotic trajectories do not appear for the overdamped pendulum. However, another possibility of non-trivial behavior, intermediate between deterministic and chaotic, does exist for the overdamped pendulum subject to two periodic fields with the following equation of motion [99],

$$\frac{d\phi}{dt} = [f\cos(\omega t) + A\cos(\beta\omega t)]\sin\phi. \tag{4.29}$$

Introducing the new variable $u(t)$,

$$u(t) \equiv \ln\tan\left(\frac{\phi(t)}{2}\right), \tag{4.30}$$

one obtains the solution of Eq. (4.29),

$$u(t) = \frac{f}{\omega}\sin(\omega t) + \frac{A}{\beta\omega}\sin(\beta\omega t). \tag{4.31}$$

Let us replace the continuous time in Eqs. (4.29)-(4.31) by the discrete times $2\pi n/\omega$ (stroboscopic plot). Then, Eq. (4.30) becomes

$$u\left(\frac{2\pi n}{\beta\omega}\right) = \frac{f}{\omega}\sin\left(\frac{2\pi n}{\beta\omega}\right) + u(0) \tag{4.32}$$

or

$$\tan\left[\frac{\phi\left(2\pi n/\beta\omega\right)}{2}\right] = \tan\left[\frac{\phi\left(0\right)}{2}\right]\exp\left[\frac{f}{\omega}\sin\left(\frac{2\pi n}{\beta\omega}\right)\right] \quad (4.33)$$

A distinction needs to be made between three cases:

1. The parameter β^{-1} is an integer. Then, $\sin\left(2\pi n/\beta\right)$ vanishes, and the pendulum comes back to its initial position $\phi\left(0\right)$ after some integer number of cycles.

2. The parameter β^{-1} is a rational number, the ratio of two integers, $\beta^{-1} = p/q$. Then, $\sin\left(2\pi np/q\right)$ vanishes after q cycles.

3. Parameter β^{-1} is an irrational number. Then, the motion is not periodic because the sin factor never vanishes. As a result of this incommensurate increase of ϕ at each stage, the motion becomes "erratic".

To better understand "erratic" motion, consider the correlation function $C\left(m\right)$ of $\tan\left[\phi\left(t\right)/2\right]$ associated with n-th and $(n+m)$-th points,

$$C\left(m\right) = \lim_{N\to\infty}\frac{1}{N}\sum_{n=0}^{N}\left\langle\tan\left[\frac{1}{2}\phi\left(\frac{2\pi n}{\beta}\right)\right]\tan\left\{\frac{1}{2}\phi\left[\frac{2\pi\left(n+m\right)}{\beta}\right]\right\}\right\rangle$$

$$= \left\langle\tan^2\left(\frac{1}{2}\phi\left(0\right)\right)\right\rangle\lim_{N\to\infty}\frac{1}{N}\sum_{n=0}^{N}$$

$$\times\exp\left\{\frac{2f}{\omega}\cos\left(\frac{\pi m}{2\beta}\right)\sin\left[\frac{\pi}{\beta}\left(n+\frac{1}{2}m\right)\right]\right\} \quad (4.34)$$

Equation (4.33) and the formula for the sum of sines have been used in the last equality in (4.34). Using the generating function for Bessel functions $I\left(z\right)$, one can rewrite (4.34) as

$$C\left(m\right) = 2\pi\left\langle\tan^2\left(\frac{\phi\left(0\right)}{2}\right)\right\rangle I_0\left(\frac{2f}{\omega}\cos\left(\frac{\pi m}{2\beta}\right)\right). \quad (4.35)$$

The correlation function $C\left(m\right)$ depends on time m (measured in units of $\pi/2\beta$), and its Fourier spectrum depends on the ratio β of the two frequencies involved. For irrational β, the Fourier spectrum of (4.35) is broadband with amplitude I_0 decreasing slowly with increasing m. This "erratic" spectrum bears a resemblance to the broadband spectrum for deterministic chaos.

4.3.3   Vibrational resonance

Analysis of Eq. (4.28) showed that a second periodic field can be used to control the chaotic behavior of the underdamped pendulum. Another interesting result is the appearance of a new type of resonance, called vibrational resonance [137]. Like stochastic resonance, vibrational resonance manifests itself in the enhancement of a weak periodic signal through a high-frequency periodic field, instead of through noise as in the case of stochastic resonance. To study small oscillations of the pendulum, we expand $\sin \phi$ and keep the first two terms in its series, which replaces the sin potential by a bistable potential. The equation of motion then has the following form,

$$\frac{d^2\phi}{dt^2} + \gamma \frac{d\phi}{dt} - \omega_0^2 \phi + \beta \phi^3 = f \sin(\omega t) + A \sin(\Omega t). \qquad (4.36)$$

The following simple explanation have been put forward for vibrational resonance [138]. Suppose that an additional field has an amplitude larger than the barrier height, $A > \omega_0^2/4\beta$, and has high frequency $\Omega >> \omega$. The former means that during each half-period, this field transfers the system from one potential well to the other. A similar situation holds in a random system where the large-amplitude field is replaced by a random force which plays the same role of switching a system between two minima. Therefore, by choosing an appropriate relation between the input signal $f \sin(\omega t)$ and the amplitude A of the large signal (or the strength of the noise), one can obtain a non-monotonic dependence of the output signal on the amplitude A (vibrational resonance) or on the noise strength (stochastic resonance).

We look for the solution of Eq. (4.36) in the form

$$\phi(t) = \psi(t) - \frac{A \sin(\Omega t)}{\Omega^2}. \qquad (4.37)$$

The first term on the right-hand side will be assumed to vary significantly only over times of order t, whereas the second term varies rapidly. Substituting (4.37) into (4.36), one can perform an averaging over a single cycle of $\sin(\Omega t)$. The slowly varying term $f \sin(\omega t)$ does not change significantly during this short time. All odd powers of $\sin(\Omega t)$ vanish upon averaging whereas the $\sin^2(\Omega t)$ term gives $\frac{1}{2}$. Finally, one obtains the following equation for $\Psi(t)$, the mean value of $\psi(t)$ during the oscillation, $\Psi(t) = < \psi(t) >$,

$$\frac{d^2\Psi}{dt^2} + \gamma \frac{d\Psi}{dt} - \left(\omega_0^2 - \frac{3\beta A^2}{2\Omega^4} \right) \Psi + \beta \Psi^3 = f \sin(\omega t). \qquad (4.38)$$

One can say that Eq. (4.38) is the "coarse-grained" version (with respect to time) of Eq. (4.36).

For $3\beta A^2/2\Omega^4 > \omega_0^2$, the phenomenon of dynamic stabilization [139] occurs, namely, the high-frequency external field transforms the previously unstable position $\Psi = 0$ into a stable position. Note that in contrast to the situation considered in Part 5 (stabilization of the inverted pendulum by the oscillations of the suspension point which appear multiplicatively in the equation of motion), here the high-frequency oscillations enter the equation of motion (4.36) additively. An approximate solution of Eq. (4.38) can be obtained [29].

The numerical solution of the original Eq. (4.36) has been performed [137] for different values of f and ω ($f = 0.025$, 0.05, 0.1 and 0.2; $\omega = 0.01$, 0.05, 0.1, 0.2 and 0.3). The graphs of the amplitude of the output signal as a function of the amplitude A of the high-frequency field have a bell shape, showing the phenomenon of vibrational resonance. For ω close to the frequency ω_0 of the free oscillations, there are two resonance peaks, whereas for smaller ω there is only one resonance peak. These different results correspond to two different oscillatory processes, jumps between two wells and oscillations inside one well.

Chapter 5

Inverted Pendulum

5.1 Oscillations of the suspension axis

A simple pendulum has two equilibrium positions, $\phi = 0$ and $\phi = \pi$, with the downward position being stable, whereas the upward position is unstable. Stability means that when being displaced from the initial position, the pendulum makes a few transient oscillations and then returns to its initial position due to friction. Without dissipation, the disturbed pendulum will oscillate forever about the stable downward position. During oscillations between the highest positions of the bob, the potential energy, which is maximal at these positions, switches back and forth to kinetic energy which is maximal at the downward position.

As we have described in Sec. 3.2, the inverted position of a pendulum can become stable for some values of the parameters [71]. There are two systematic ways to achieve this aim, either by the periodic or random vertical oscillations of the suspension point, or by the spring pendulum where one inserts a spring into a rigid rod, or by using both these means together. The inverted pendulum is very sensitive part of control devices that have many technological applications.

In Sec. 3.6.1, we discussed the general properties of the pendulum with a vertically oscillating suspension point, described by the equation,

$$\frac{d^2\phi}{dt^2} + \frac{1}{l}\left[g + a\omega^2 \sin(\omega t)\right]\sin\phi = 0. \tag{5.1}$$

We now explain physically [6] why rapid vertical oscillations of the suspension point make stable the inverted (upward) position of a pendulum. From the two torques in Eq. (5.1), the torque of the gravitational force tends to tip the pendulum downward while the torque of the inertial force (averaged over the period ω of rapid oscillations) tends to return the pendulum to the inverted position. If the latter is large enough, the inverted position of a pendulum will be stable. The quantitative criterion can be obtained in the following way [140]. One starts with the ansatz that the angle ϕ is the superposition of two components

$$\phi = \phi_{slow} + \phi_{fast}, \tag{5.2}$$

implying

$$\sin \phi \approx \sin \phi_{slow} + \phi_{fast} \cos \phi_{slow} \tag{5.3}$$

where the "slow" angle, ϕ_{slow}, has a small variation during a period of constrained oscillations, whereas the "fast" angle, ϕ_{fast}, is small with zero mean value, $\langle \phi_{fast} \rangle = 0$. The angle ϕ_{fast} oscillates with high frequency ω and has an amplitude proportional to the sine of the momentary value of $\phi = \phi_{slow}$,

$$\phi_{fast} = -\frac{u(t)}{l} \sin \phi_{slow} = -\frac{a}{l} \sin \phi_{slow} \sin (\omega t). \tag{5.4}$$

The second equation means that the average value of the gravitational torque $\langle mgl \sin \phi \rangle = \langle mgl \sin \phi_{slow} \rangle$ is the same as for a pendulum with a fixed suspension point, whereas the average value of the inertial torque

$$\langle ma\omega^2 l \sin \phi \sin (\omega t) \rangle = \langle ma\omega^2 l \sin(\phi_{slow} + \phi_{fast}) \sin (\omega t) \rangle$$
$$\approx \langle ma\omega^2 l \sin \phi_{slow} + ma\omega^2 l \phi_{fast} \cos \phi_{slow} \sin (\omega t) \rangle \tag{5.5}$$

obtains an additional term which is equal to $\frac{1}{2} ma^2 \omega^2 \sin \phi_{slow} \cos \phi_{slow}$, where Eq. (5.3) and $\overline{\sin^2 (\omega t)} = 1/2$ have been used. Comparing the latter term with the gravitational torque, one concludes that the inertial torque can exceed the gravitational torque and cause the pendulum to tip up when the following condition is fulfilled,

$$a^2 \omega^2 > 2gl. \tag{5.6}$$

The latter formula can be rewritten for the dimensionless amplitude and frequency as

$$\frac{1}{2} \frac{a^2}{l^2} \frac{\omega^2}{\omega_0^2} > 1. \tag{5.7}$$

Adding damping to Eq. (5.1) causes the stability of the inverted state to decrease while the stability of the downward position is enhanced [141].

5.2 The tilted parametric pendulum

In the preceding section, we considered the pendulum with vertical oscillations of the suspension point. The more general case involves the oscillatory displacement of the suspension point $A\sin(\Omega t)$ at an arbitrary angle ψ from the upward vertical (Fig. 5.1). Vertical oscillations correspond to the special case $\psi = 0$, whereas for horizontal oscillations $\psi = \pi/2$.

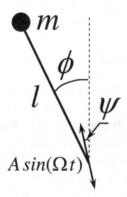

Fig. 5.1 Tilted parametric pendulum.

The equation of motion for this case can be obtained in the same way as was done for the horizontal oscillations of the suspension axis in Sec. 3.6.2. The coordinates x, z of the bob are

$$x = l\sin\phi + f\cos(\omega t)\sin\psi; \qquad z = l\cos\phi + f\cos(\omega t)\cos\psi. \qquad (5.8)$$

The kinetic energy T and the potential energy U in terms of the coordinates x and z are

$$T = \frac{1}{2}m\left[\left(\frac{dx}{dt}\right)^2 + \left(\frac{dz}{dt}\right)^2\right]$$

$$= \frac{1}{2}m\left[l^2\left(\frac{d\phi}{dt}\right)^2 - 2f\omega l\sin\omega t\,(\cos\phi\sin\psi - \sin\phi\cos\psi) + f^2\omega^2\sin^2(\omega t)\right];$$

$$U = -mgl\cos\phi - mgf\cos(\omega t)\cos\psi.$$

$$(5.9)$$

Substituting (5.9) into the Lagrange function $L = T - U$, and using the Lagrange equations, one obtains, after adding the damping term,

$$\frac{d^2\phi}{dt^2} + \gamma \frac{d\phi}{dt} - \frac{g}{l} \sin \phi - \frac{f\omega^2}{l} \cos(\omega t) \sin(\phi - \psi) = 0. \qquad (5.10)$$

For $\psi = 0$ and $\psi = \pi/2$, Eq. (5.10) reduces to Eqs. (3.31) and (3.47) for vertical and horizontal oscillations of the suspension axis, respectively, as required.

Introducing the dimensionless time $\hat{\tau} = \omega t$, damping $\gamma/\omega = \hat{\gamma}$, frequency $\hat{\omega} = \omega^{-1}\sqrt{g/l}$ and amplitude $\hat{f} = f/l$ into (5.10) yields

$$\frac{d^2\phi}{d\hat{\tau}^2} + \hat{\gamma} \frac{d\phi}{d\hat{\tau}} - \hat{\omega}^2 \sin \phi - \hat{f} \cos(\hat{\tau}) \sin(\phi - \psi) = 0. \qquad (5.11)$$

We use here two slightly different methods of analysis of Eq. (5.11). First, just as in the previous section, we may divide the motion into fast and slow, according to Eq. (5.2) and, using the effective potential method, as in that section, one concludes [142] that result depends on the value of parameter Ω, which is equal to the squared ratio of the frequency ω of the external field to the critical angular velocity $\omega_{cr} = \sqrt{gl/f^2}$,

$$\Omega = \frac{\omega^2 f^2}{gl}. \qquad (5.12)$$

The interesting question is whether there are stable oscillations around the angle ϕ_0 for different tilted angle ψ. An approximate analytical calculation [142] gives the following values of the equilibrium angle ϕ_0 for given ψ,

$$\sin \phi_0 + \frac{\Omega}{2} \sin[2(\phi_0 - \psi)] = 0. \qquad (5.13)$$

For $1 < \Omega < 2$, there are stable oscillations near the vertical upward position although some angles in the range $0 < \psi < \pi/2$, are not stable for $|\psi - \phi_0| < \pi/2$. For $\Omega \geq 2$, there are stable oscillations for $|\psi - \phi_0| < \pi/2$ for all ψ. Finally, for $\Omega \leq 1$, there are no stable oscillations around the vertical upward position, although for all $\pi/2 < \psi < \pi$, there are stable oscillations with $\phi_0 < \pi$.

Another method of analysis of Eq. (5.11), already used in Sec. 4.2.4, is the application of the method of multiplicative time scales by introducing three time scales [143] by scaling the parameters in (5.11) according to

$$\hat{\omega}^2 = \epsilon^2 \omega^2; \qquad \hat{f} = \epsilon f; \qquad \hat{\gamma} = \epsilon \mu \qquad (5.14)$$

with $\epsilon \ll 1$. We seek a solution of Eq. (5.11) of the form

$$\phi(t, \epsilon) = \phi_0(t_0, t_1, t_2) + \epsilon \phi_1(t_0, t_1, t_2) + \epsilon^2 \phi_2(t_0, t_1, t_2) + \cdots \quad (5.15)$$

where $t_0 = \hat{\tau}$, $t_1 = \epsilon \hat{\tau}$, and $t_2 = \epsilon^2 \hat{\tau}$. Substituting Eq. (5.15) into Eq. (5.11) and equating coefficients of powers of ϵ yields

$$D_0^2 \phi_0 = 0, \quad (5.16)$$

$$D_0^2 \phi_1 + 2D_0 D_1 \phi_0 + \mu D_0 \phi_0 + f \sin(\phi_0 - \psi) \sin(t_0) = 0, \quad (5.17)$$

$$D_0^2 \phi_2 + 2D_0 D_1 \phi_1 + 2D_0 D_2 \phi_{0-} + D_1^2 \phi_0 + \mu(D_0 \phi_1 + D_1 \phi_0)$$
$$+ \omega^2 \sin \phi_0 + f \cos(\phi_0 - \psi) \phi_1 \sin(\tau) = 0 \quad (5.18)$$

where $D_n = d/dt^n$. The general solution of Eq. (5.16) is

$$\phi_0 = c_1(t_1, t_2) t_0 + c_0(t_1, t_2). \quad (5.19)$$

Substituting Eq. (5.19) into Eq. (5.17) yields

$$D_0^2 \phi_1 = -f \sin(\phi_0 - \psi) \sin(t_0). \quad (5.20)$$

The particular solution of Eq. (5.20) is

$$\phi_1 = f \sin(\phi_0 - \psi) \sin(t_0). \quad (5.21)$$

Substituting Eq. (5.21) into Eq. (5.18) gives

$$D_0^2 \phi_2 = -D_1^2 \phi_0 - \mu D_1 \phi_0 + \omega^2 \sin \phi_0 - \frac{1}{4} f^2 \sin 2(\phi_0 - \psi)$$
$$+ \left[2f \cos(\phi_0 - \psi) D_1 \phi_0 + \mu f \sin(\phi_0 - \psi) \sin(t_0) + \frac{1}{4} f^2 \sin 2(\phi_0 - \psi) \sin 2t_0 \right]. \quad (5.22)$$

In the method of multiple time scales, one reduces all secular terms to zero. In our case, these are terms that do not contain $\sin t_0$ or $\sin 2t_0$,

$$D_1^2 \phi_0 + \mu D_1 \phi_0 - \omega^2 \sin \phi_0 + \frac{1}{4} f^2 \sin 2(\phi_0 - \psi) = 0. \quad (5.23)$$

Multiplying (5.23) by ϵ^2, leads to the following equation,

$$\frac{d^2 \phi}{dt^2} + \gamma \frac{d\phi}{dt} - \hat{\omega}^2 \sin \phi + \frac{1}{4} \hat{f}^2 \sin 2(\phi - \psi) = 0. \quad (5.24)$$

With the help of the method of multiple time scales, Eq. (5.11) reduces to the much simpler Eq. (5.24). The stability of the equilibrium states $(d/dt^n = 0)$ can be easily analyzed [143], leading to the following results:

1. The stable equilibrium state depends on the amplitude of an external field. For large amplitudes, the stable state approaches the excitation direction ($\phi = \psi$). With decreasing amplitude, the stable state is located at angles larger than ψ.

2. There is a critical frequency, $\omega_c = 0.093\sqrt{g/l}$, such that for $\omega \lesssim \omega_c$, the upward position of the pendulum is unstable. The stable position is located in the regime $\psi <| \phi |< \pi/2$, which differs from both the upward and the titled direction. With increasing frequency, the inclination of the pendulum in the stable equilibrium approaches the excitation direction $\phi = \psi$.

Experiments [143] are in qualitative agreement with these theoretical results.

5.3 Random vibrations of the suspension axis

The inverted pendulum becomes stable not only through oscillations of the suspension axis, but also through its random vibrations. The latter case is described by the following equation,

$$\frac{d^2\phi}{dt^2} + \gamma\frac{d\phi}{dt} + \left[1 + \omega^2\xi\left(t\right)\right]\sin\phi = 0, \qquad (5.25)$$

where ω^2 is the acceleration of the suspension axis in terms of $\omega_0 = \sqrt{g/l}$, and the random noise $\xi\left(t\right)$ has been chosen [145] in the form of narrowband ($1 << \lambda << \Omega$) colored noise with a correlator of the form (1.53). For a non-resonant random force, the fluctuations $\delta\phi$ caused by the random force are small, $\delta\phi << \langle\phi\rangle$. Substituting $\phi = \langle\phi\rangle + \delta\phi$ in (5.25) leads to the following two equations,

$$\frac{d^2\langle\phi\rangle}{dt^2} + \gamma\frac{d\langle\phi\rangle}{dt} + \sin\langle\phi\rangle + \omega^2\cos\langle\phi\rangle\langle\xi\left(t\right)\delta\phi\rangle = 0,$$
$$\frac{d^2\delta\phi}{dt^2} + \gamma\frac{d\delta\phi}{dt} + \cos\langle\phi\rangle\,\delta\phi + \omega^2\xi\left(t\right)\sin\langle\phi\rangle = 0. \qquad (5.26)$$

The steady-state solutions of Eqs. (5.26), $\langle\phi\rangle = \pi$, $\delta\phi = 0$, correspond to the inverted position of the pendulum. One may investigate the stability of these solutions by linearizing Eq. (5.26) with respect to small deviations $\psi = \langle\phi\rangle - \pi$ and $\delta\phi$, which gives

$$\frac{d^2\psi}{dt^2} + \gamma\frac{d\psi}{dt} - \psi - \omega^2\langle\xi\left(t\right)\delta\phi\rangle = 0, \qquad (5.27)$$

$$\frac{d^2\delta\phi}{dt^2} + \gamma\frac{d\delta\phi}{dt} - \delta\phi - \omega^2\xi(t)\psi = 0. \tag{5.28}$$

Inserting the steady-state solution of Eq. (5.28) into (5.27), using (1.53) and the conditions $\Omega \gg 1, \gamma, \lambda$, leads to

$$\frac{d^2\psi}{dt^2} + \gamma\frac{d\psi}{dt} + \left(\Omega^2\sigma^2 - 1\right)\psi = 0. \tag{5.29}$$

It follows from this equation that the mean deviation of the pendulum from its equilibrium inverted position will decay, and the inverted pendulum will be stable if

$$\Omega^2\sigma^2 > 1. \tag{5.30}$$

The latter condition remains the stability condition (5.7) of the pendulum with periodic oscillations of the suspension axis.

New phenomena appear if one introduces additive white noise $\eta(t)$ into Eq. (5.25),

$$\frac{d^2\phi}{dt^2} + \gamma\frac{d\phi}{dt} + \left[1 + \omega^2\xi(t)\right]\sin\phi = \eta(t). \tag{5.31}$$

In the presence of additive noise $\eta(t)$, an additional maximum and minimum appear in the probability distribution function $P(\phi, d\phi/dt)$ described by the Fokker-Planck equation. To see this, let us again use the "slow" variable $\langle\phi\rangle$ and the "fast" variable $\delta\phi$. The narrow-band random process $\xi(t)$ can be represented by the "slow" variables $\xi_1(t)$ and $\xi_2(t)$, defined by

$$\xi(t) = \xi_1(t)\cos(\Omega t) + \xi_2(t)\sin(\Omega t). \tag{5.32}$$

Similarly, one can write $\delta\phi$ through the "slow" variables $A(t)$ and $B(t)$,

$$\delta\phi = A(t)\cos(\Omega t) + B(t)\sin(\Omega t). \tag{5.33}$$

Substituting $\phi = \langle\phi\rangle + \delta\phi$, inserting Eqs. (5.32) and (5.33) into (5.31), and equating the slowly varying component and the coefficients of $\cos(\Omega t)$ and $\sin(\Omega t)$, one obtains [145] equations for $\langle\phi\rangle$, A and B. For $\Omega \gg 1, \gamma$, one finds

$$A \approx \xi_1\sin\langle\phi\rangle; \qquad B \approx \xi_2\sin\langle\phi\rangle \tag{5.34}$$

and the following equation for $\langle\phi\rangle$,

$$\frac{d^2\langle\phi\rangle}{dt^2} + \gamma\frac{d\langle\phi\rangle}{dt} + \sin\langle\phi\rangle + \frac{\omega^2\sin 2\langle\phi\rangle}{2}\sigma^2 = \eta(t). \tag{5.35}$$

The steady-state solution of the Fokker-Planck equation, corresponding to the Langevin equation (5.31), has the following form [145]

$$P\left(\langle\phi\rangle, d\langle\phi\rangle/dt\right) = N\exp\left\{\frac{1}{D}\left(\frac{d\langle\phi\rangle}{dt}\right)^2 + \left[\frac{2}{D}\cos\langle\phi\rangle + \frac{\Omega^2\sigma^2}{2D}\cos 2\langle\phi\rangle\right]\right\}$$

(5.36)

where D is the strength of the white noise and N is the normalization constant. For $\Omega^2\sigma^2 > 1$, the function $P(\phi, d\phi/dt)$ has two maxima (for $\langle\phi\rangle = 0$ and $\langle\phi\rangle = \pi$) and one minimum (for $\langle\phi\rangle = \arccos(-1/\Omega^2\sigma^2)$), whereas for $\Omega^2\sigma^2 < 1$, there is one maximum (at $\langle\phi\rangle = 0$) and one minimum (at $\langle\phi\rangle = \pi$). Thus, random vibrations of the suspension axis causes metastability, in addition to stabilizing the inverted position of the pendulum.

5.4 Spring pendulum

As we have seen in Sec. 5.1, when condition (5.6) is satisfied, the oscillation of the suspension point makes the upward position stable. Another way to accomplish this is to replace the rigid pendulum by a spring, as discussed in Sec. 3.7. A new reservoir of energy then appears, that of a stretched, but not bent spring. Just as for the kinetic-potential energy transition in a simple pendulum, during the oscillations of a spring pendulum, the elastic energy of the spring switches back and force to the mechanical energy of oscillations [146].

A spring pendulum becomes unstable when the driving force X^2 in (3.73), of frequency $2\hat{\omega}_0$, is of order of the characteristic frequency ω_s. This becomes obvious in the case of dominant spring mode $X << Z$ [147]. Neglecting the nonlinear term in (3.73), one obtains

$$Z = a\cos\left(\omega_s t\right).$$

(5.37)

Substituting (5.37) into (3.72) produces the Mathieu equation

$$\frac{d^2X}{dt^2} + \left[\hat{\omega}_0^2 - a\frac{\hat{\omega}_0^2 - \omega_s^2}{l}\cos\left(\omega_s t\right)\right]X = 0.$$

(5.38)

When $\omega_s/\hat{\omega}_0 = 2/n$, where $n = 1, 2...$, the solutions of Eq. (5.38) are unstable even for small values of a, Since $\omega_s > \hat{\omega}_0$, $n = 1$ and the instability occurs at

$$\omega_s = 2\hat{\omega}_0.$$

(5.39)

The occurrence of instability can be also seen from Eq. (3.73), when one takes into account that a resonance type of coupling occurs when X^2 has spring velocity ω_s. This will be the case when condition (5.39) is satisfied since squaring a sinusoidal function doubles its frequency.

Since the pendulum is a conservative system, the resonant growth of one mode occurs at the expense of the other mode. Therefore, the energy transfers back and forth between these two modes (autoparametric resonance). It follows that when the spring frequency $\omega_s^2 = \kappa/m$ is about twice the pendulum frequency $\hat{\omega}_0^2 = g/l$, a spring pendulum performs periodic oscillations about its upper vertical position. Condition (5.39) implies that the spring pendulum oscillates in such a way that the spring goes up and down twice during one oscillation of the pendulum.

5.5   Spring pendulum driven by a periodic force

This analysis is a generalization of the analysis considered in preceding sections of a pendulum with the oscillating suspension point and a spring pendulum. The analytic solution of the linear problem and numerical simulation of the non-linear equations yield new results, such as a new type of resonance between the frequency of a spring and that of an external force, and the appearance of limit cycle oscillations near the upper position of the pendulum.

It is convenient to rewrite the Lagrangian (3.63) in polar coordinates R and ϕ, defined as $z = R\cos\phi$, $x = R\sin\phi$,

$$\left(\frac{dx}{dt}\right)^2 + \left(\frac{dz}{dt}\right)^2 = \left(\frac{dR}{dt}\right)^2 + \left(R\frac{d\phi}{dt}\right)^2 \tag{5.40}$$

and

$$L = \frac{m}{2}\left[\left(\frac{dR}{dt}\right)^2 + \left(R\frac{d\phi}{dt}\right)^2\right] - m\left(g + \frac{d^2u}{dt^2}\right)R\cos\phi - \frac{\kappa}{2}(R - l_0)^2 \tag{5.41}$$

with $R = \sqrt{X^2 + (Z + l_0 + g/\omega_s^2)^2}$.

According to Eq. (5.41), the suspension point performs vertical oscillations $u(t) = a\cos(\Omega t)$. The suspension point has acceleration d^2u/dt^2 relative to our inertial frame of reference. We introduce a non-inertial frame which itself has this acceleration. Therefore, in this non-inertial frame, gravity g is replaced by $g + d^2u/dt^2$. Such is indeed the case in Eq. (5.41).

When the bob starts to oscillate near the downward position, one easily obtains [148] from (5.41) the following equations of motion for $\hat{r}$ and $\hat{\phi} = \pi - \phi$,

$$\frac{d^2\hat{r}}{dt^2} + \left[\omega_s^2 - \left(\frac{d\phi}{dt}\right)^2\right]\hat{r} = \left(1 + \frac{\omega_0^2}{\omega_s^2}\right)\left(\frac{d\phi}{dt}\right)^2 - \omega_0^2(1 - \cos\phi) + \frac{d^2\hat{u}}{dt^2}\cos\phi,$$

(5.42)

$$\frac{d^2\hat{\phi}}{dt^2} + \frac{2}{1 + \omega_0^2/\omega_s^2 + \hat{r}}\frac{d\hat{r}}{dt}\frac{d\hat{\phi}}{dt} + \frac{1}{1 + \omega_0^2/\omega_s^2 + \hat{r}}\left(\omega_0^2 + \frac{d^2\hat{u}}{dt^2}\right)\sin\hat{\phi} = 0,$$

(5.43)

where $\hat{u}(t) = u(t)/l_0$, and the dimensionless relative elongation $\hat{r}$ of the pendulum is given by

$$\hat{r} = \frac{R}{l_0} - \left(1 + \frac{\omega_0^2}{\omega_s^2}\right).$$

(5.44)

When the bob starts to oscillate near the upper vertical position, one obtains the following equations of motion for r and ϕ,

$$\frac{d^2r}{dt^2} + \left[\omega_s^2 - \left(\frac{d\phi}{dt}\right)^2\right]r = \left(1 - \frac{\omega_0^2}{\omega_s^2}\right)\left(\frac{d\phi}{dt}\right)^2 + \omega_0^2(1 - \cos\phi) - \frac{d^2\hat{u}}{dt^2}\cos\phi,$$

(5.45)

$$\frac{d^2\phi}{dt^2} + \frac{2}{1 - \omega_0^2/\omega_s^2 + r}\frac{dr}{dt}\frac{d\phi}{dt} - \frac{1}{1 - \omega_0^2/\omega_s^2 + r}\left(\omega_0^2 + \frac{d^2\hat{u}}{dt^2}\right)\sin\phi = 0,$$

(5.46)

where the dimensionless relative elongation r of the pendulum is given by

$$r = \frac{R}{l_0} - \left(1 - \frac{\omega_0^2}{\omega_s^2}\right).$$

(5.47)

As expected, the equations of motion near the upper position, (5.45) and (5.46), differ from the equations near the downward position, (5.42) and (5.43), by the signs of ω_0^2 and $\hat{u}$.

Note that Eqs. (5.45) and (5.46) are written for the relative elongation r, and not for the polar coordinate R [149]. When rewritten in terms of R, these equations take the standard form for $\hat{u} = 0$.

The linearized equation (5.46) with $\hat{u} = (a/l_0)\cos(\Omega t)$, written in dimensionless time $\tau = \Omega t$, is the Mathieu equation,

$$\frac{d^2\phi}{d\tau^2} - \frac{1}{1 - \omega_0^2/\omega_s^2}\left[\frac{\omega_0^2}{\Omega^2} - \frac{a}{l_0}\cos(\tau)\right]\phi = 0.$$

(5.48)

There exists a comprehensive literature concerning Mathieu equations of the general form

$$\frac{d^2\phi}{dt^2} + \gamma\frac{d\phi}{dt} + [\alpha + \beta\cos(t)]\phi = 0. \tag{5.49}$$

For small amplitude β of an external field, the first zone of stability is bordered from below by the curve [150; 151]

$$\alpha = -0.5\beta^2. \tag{5.50}$$

In terms of the parameters in (5.48), this curve is

$$a^2\Omega^2 = 2l_0^2\omega_0^2\left(1 - \omega_0^2/\omega_s^2\right) = 2gl_0\left(1 - \omega_0^2/\omega_s^2\right) \tag{5.51}$$

which, for the rigid pendulum ($\omega_s = \infty$), coincides with the curve obtained in Sec. 5.1 by a different method. For the upper border of the stability region, one has [150; 151], $\alpha = 0.25 - 0.556\beta$, which corresponds to

$$\frac{\Omega^2}{\omega_0^2} = \left[0.556\frac{a}{l_0} - 0.25\left(1 - \frac{\omega_0^2}{\omega_s^2}\right)\right]. \tag{5.52}$$

The inverted pendulum has vertical oscillations of the suspension point, $u = A\cos(\Omega t)$. These oscillations cause the upward position to become stable if the following condition is fulfilled,

$$A^2\Omega^2 > 2gl \equiv 2\omega_0^2l_0l. \tag{5.53}$$

Due to elastic stress and gravity, the length l of a spring pendulum in the upward position decreases from its initial length l_0. According to Eq. (3.59), $l = l_0\left(1 - \omega_0^2/\omega_s^2\right)$. Substituting into (5.53), one can rewrite the boundary of stability of the upward position of a pendulum,

$$\frac{A^2}{l_0^2} = 2\frac{\omega_0^2}{\Omega^2}\left(1 - \frac{\omega_0^2}{\omega_s^2}\right). \tag{5.54}$$

It is clear that the amplitude of the external oscillations A cannot be larger than the length l of a pendulum, which means that for the maximum possible amplitude $A = l_0$, Eq. (5.54) gives $\Omega^2 > 2\omega_0^2/\left(1 - \omega_0^2/\omega_s^2\right)$.

The spring results in new effects, in addition to the transition from (5.53) to (5.54). The following two new effects have not yet been sufficiently explored.

1. If the characteristic frequency of driving force Ω is comparable to the characteristic frequency of the spring, ω_s, resonance phenomena take place (especially for small ϕ), and the pendulum becomes unstable. The

impact of this resonance on the lower stability-instability curve is expressed by the replacement of (5.53) by (5.54). However, its influence on the upper stability-instability curve is much more appreciable. Indeed, according to the linear analysis of the Mathieu equation, these two curves, described by Eqs (5.51) and (5.52), have opposite slope at small A/l_0. In that case, the upper curve crosses the $\Omega - \omega_s$ resonance curve, and the system becomes unstable. Hence, we expect that the nonlinear analysis will produce an upper stability-instability curve substantially different from the linear result (5.52).

2. Comparison of Eqs. (5.42) and (5.45) shows that in the linear approximation, the time dependence of the elongation $r(t)$ is determined by an external field, and the influence of the angular mode for small ϕ is described by the term $r(d\phi/dt)^2$. The latter means that for small $\phi(t = 0)$, say, only 10^{-4}, the appropriate $r(t)$ not need be small. On the other hand, an external field enters Eq. (5.48) multiplicatively, thereby exerting much weaker control over $\phi(t)$ than over $r(t)$. Hence, a small initial $\phi(t = 0)$ will remain small for all t.

We conclude that the above linear stability analysis based on the Mathieu equations and power-type coupling of oscillatory modes is not complete, and one must obtain the numerical solution of the full nonlinear equations. Such a numerical analysis was performed using the MathCad program with the help of the built-in function "differential equation solver". For this purpose, the system of two equations of motion, (5.45) and (5.46), was transformed into a system of four first-order differential equations,

$$\dot{r} = P_r,$$

$$\dot{P_r} = -\omega_s^2 r + \left(1 - \omega_0^2/\omega_s^2 + r\right) P_\theta^2 + \omega_0^2 (1 - \cos\phi) - \left(d^2\hat{u}/dt^2\right)\cos\phi,$$

$$\dot{\phi} = P_\phi,$$
(5.55)

$$\dot{P_\phi} = -\left(1 - \omega_0^2/\omega_s^2 + r\right)^{-1}\left[2P_r P_\phi - \left(\omega_0^2 + d^2\hat{u}/dt^2\right)\sin\phi\right].$$

One can draw the dimensionless stable-unstable phase diagram in $\{\Omega/\omega_s, A/l_0\}$ variables (for constant ω_0), showing that the stability region is bounded by three straight lines and vanishes at large amplitudes A (see Fig. 5.2(a)).

An additional cut-off appears at small A, where the parabola $\Omega \approx A^{-1}$ approaches the horizontal line $\Omega/\omega_s = 1$. The latter condition implies the appearance of a resonance when the frequency Ω of the driving force is equal to eigenfrequency ω_s of the spring, leading to a loss of stability. Therefore,

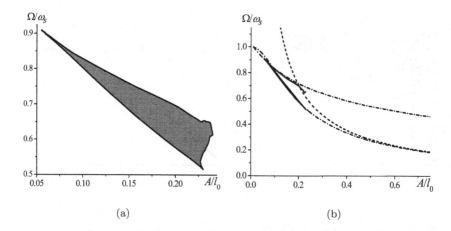

(a)   (b)

Fig. 5.2   (a) Dimensionless stable-unstable phase diagram in $\{\Omega/\omega_s, A/l_0\}$ variables for constant $\omega_0/\omega_s = 0.1$, obtained from numerical solutions of the full nonlinear equation system (5.55). The hatched region corresponds to stable oscillations of the inverted pendulum; (b) Dashed-dot lines are the instability-stability curves described by empirical Eqs. (5.56) (lower curve) and (5.54) (upper curve) which closely coincide to the top of the lower and upper boundary of the stability region, and the dashed lines describe the lower instability-stability boundary of the rigid pendulum with the small corrections given by (5.54).

the parabola $\Omega \approx A^{-1}$ does not go to infinity, as was the case for a rigid pendulum ($\omega_s = \infty$), but rather approaches the line $\Omega = \omega_s$. In this transition region, the boundary of stability (5.54) can be replaced by the empirical formula

$$\frac{A^2}{l_0^2} = 2\frac{\omega_0^2}{\omega_s^2}\left(1 - \frac{\omega_0^2}{\omega_s^2}\right)\left(\frac{\omega_s^2}{\Omega^2} - 1\right), \tag{5.56}$$

which reduces to (5.54) for small Ω/ω_s, and describes the resonance $\Omega \simeq \omega_s$ for small A. The difference (for small A) and congruence (for large A) between the curves described by Eqs. (5.54) and (5.56) are shown in Fig. 5.2b.

Analogously, the dimensionless stable-unstable phase diagrams can be drawn in the variables $\{\Omega/\omega_s, \omega_0/\omega_s\}$ (Fig. 5.3).

The lower lines, defining the stability region on these diagrams, are accurately described by Eq. (5.56). The congruence (for $\omega_0 \ll \omega_s$) and difference (for larger $\omega_0 < \omega_s$) between the curves described by Eqs. (5.54) and (5.56) are shown in Fig. 5.3(b).

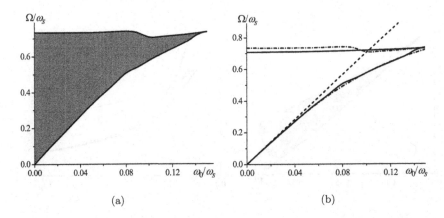

Fig. 5.3   Same as for Fig. 5.2, but in $\{\Omega/\omega_s, \omega_0/\omega_s\}$ variables for $A/l_0 = 0.2$.

The numerical solutions of the full nonlinear Eqs. (5.55) exhibit some special features that cannot be obtained from the linear analysis. The latter give lower and upper lines defining the stability region, whereas the former show some "fine structure" of the stable solutions. For small fixed A/l_0, say, $A/l_0 = 0.1$, upon increasing Ω/ω_s, one obtains only unstable solutions for $\Omega/\omega_s < 0.802$. At $\Omega/\omega_s = 0.802$, one crosses the lower stable-unstable line passing to stable solutions, and finally, for $\Omega/\omega_s = 0.829$, one returns to an unstable region. However, there are different types of stable solutions in the regime $0.802 < \Omega/\omega_s < 0.829$. For initial displacement $\phi_0 = 10^{-4}$, the oscillations have amplitude of the same order for $0.802 < \Omega/\omega_s < 0.826$ (see Fig. 5.5, where $\Omega/\omega_s = 0.810$), whereas for $0.826 < \Omega/\omega_s < 0.829$, the same initial condition leads to limit cycles oscillations of large amplitude (see Fig. 5.6, with $\Omega/\omega_s = 0.828$). The analogous situation occurs for $A/l_0 = 0.2$. For initial displacement $\phi_0 = 10^{-4}$, the oscillations have amplitude of the same order for $0.576 < \Omega/\omega_s < 0.696$, whereas for $0.696 < \Omega/\omega_s < 0.699$, the same initial condition leads to limit-cycles oscillations of large amplitude. At $A/l_0 = 0.23$, the stability regime ($0.563 < \Omega/\omega_s < 0.654$) is divided into three parts: for initial displacement $\phi_0 = 10^{-4}$ and for $0.578 < \Omega/\omega_s < 0.652$, the oscillations have amplitudes of the same order, whereas for $0.563 < \Omega/\omega_s < 0.578$ and for $0.652 < \Omega/\omega_s < 0.654$, the same initial condition leads to limit-cycles oscillations of large amplitude. And at $A/l_0 = 0.23$, the small initial displacement $\phi_0 = 10^{-4}$ results in limit-cycles oscillations of large amplitude throughout the stability regime, $0.587 < \Omega/\omega_s < 0.659$.

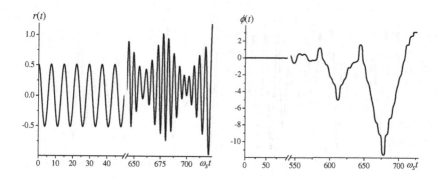

Fig. 5.4 Time dependence of the radial, $r(t)$, and angular, $\phi(t)$, coordinates for $\Omega/\omega_0 = 0.848$, at $\phi(t=0) = 10^{-4}$ and $r(t) = 0$ for $A/l_0 = 0.1$, showing the instability of $\phi(t)$.

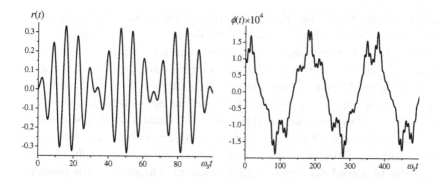

Fig. 5.5 Small oscillations which are the stable solutions of the nonlinear equations of motion for an initial disturbance $\phi(t=0) = 10^{-4}$ and $r(t) = 0$, for $A/l_0 = 0.1$, $\Omega/\omega_s = 0.846$.

The inclusion of damping probably leads to the disappearance of small oscillations and a shift of the limit cycles. For larger A/l_0, this phenomenon — alternation of small oscillations and limit cycles — occurs many times in the stable regime of the parameters. Note that such oscillations have been found for the rigid pendulum for non-small initial disturbances [152]. Analogous to Fig. 5.3, the $\{\Omega/\omega_s, \omega_0/\omega_s\}$ dependence is described by a straight line that changes its form close to resonance, $\Omega = \omega_s$, according to Eq. (5.56).

Thus far, we have considered the undamped Mathieu equation. The

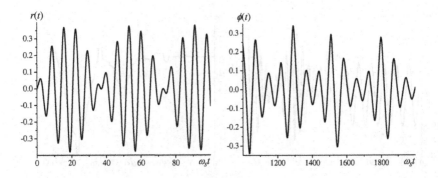

Fig. 5.6   Stable limit-cycle oscillations which are the stable solutions of the nonlinear
equations of motion for an initial disturbance $\phi\,(t = 0) = 10^{-4}$ and $r(t) = 0$, for $A/l_0 =$
0.1, $\Omega/\omega_s = 0.847$.

qualitative influence of the damping term $\gamma d\phi/dt$ in the Mathieu equation (5.49) is the following. The amplitude-frequency phase diagram of an
external field for the undamped Mathieu equation has alternating stability-
instability regimes down to zero [150]. This means that by changing the frequency of the field, the transitions between stability and instability regimes
occurs for even infinitesimal driving values of the amplitude. However, in
the presence of damping, the stability-instability boundaries do not extend
to the frequency axis. This implies that destabilization of the n-th mode
will only occur starting from a non-zero value of the driving amplitude (in
fact, β has to exceed $\beta_{cr} \approx \gamma^{1/n}$) [150]. Numerical analysis of the impact
of damping has been carried out [52] for $\omega_s = \infty$.

Chapter 6

Conclusions

Just as the harmonic oscillator is the simplest linear model, so the pendulum is one of the simplest nonlinear models. Both these models are widely used for the description of different phenomena in physics, chemistry, biology, economics and sociology. Since the majority of phenomena in Nature are nonlinear, using nonlinear models is most appropriate, although the price we have to pay for using these models is their complexity. As an illustration of this complexity, consider the response of a system to an external periodic force. The well-known resonance in linear systems implies the unbounded increase, say, of the amplitude of the harmonic oscillator, when the frequency of an external field is equal to the characteristic frequency of the oscillator. On the other hand, when such a situation occurs for a nonlinear system, the amplitude of the pendulum oscillations starts to increase, but the synchronization of frequencies breaks down since, in contrast to linear systems, the frequency of a nonlinear system depends on the amplitude of its motion. Another difference of fundamental importance between linear and nonlinear systems is the possibility of deterministic chaos in nonlinear systems, which shows the deep relationship between determinism and stochasticity, which, like relativity and quantum mechanics, goes far beyond the scope of science and forms an important part of our world outlook. Although pendulum dynamics is described by deterministic equations, deterministic chaos influences the motion of the pendulum due to an exponential change with time upon a very small change in the initial conditions. In any experiment, the initial conditions are known only to restricted accuracy and, therefore, "deterministic chaos" manifests itself in a deterministic equation. In this way, chaotic solutions show "randomlike" behavior.

The equation of motion (1.1) of the simple mathematical pendulum has periodic solutions with the period depending on the initial conditions. Only for the simple case of small oscillations, $\sin\phi \approx \phi$, does the linearized equation have a single sinusoidal solution. Adding the periodic external force to the pendulum equation leads to the appearance of both regular and chaotic solutions. The values of the damping and the constant torque influence the choice between these two types of solutions, as described in the appropriate sections of this book. All phenomena in Nature are subject to random perturbations which are described by adding a random force to the deterministic equations. Such forces may be of internal origin, such as thermal noise at non-zero temperatures, or the source of (external) noise may be due to random changes in the surrounded media. Accordingly, these sources of noise appear in additive or multiplicative form. The influence of noise on pendulum dynamics has been discussed in this book. As a rule, these sources of noises are independent, although sometimes one has to consider their correlation if they have a common source, or if the external noise is so strong that it changes the structure of the system and, hence, influences internal noise. The addition of random internal or external noise has a different influence on the regular and the chaotic solutions. The different influences of the internal and external noises may be illustrated on the joint action of deterministic chaos and noise. When deterministic chaos gives rise to chaotic trajectories, only the external noise is able to transform these trajectories from chaotic to regular.

Neglect of inertial effects facilitates the analysis of the pendulum equation. Under certain conditions, an overdamped pendulum is able to derive energy from the source of noise and transfer it into deterministic motion without any external driving force. The second law of thermodynamics forbids such processes in equilibrium systems, but, as we have shown, they may occur when the potential energy or noise are asymmetric, or when a correlation exists between additive and multiplicative noise.

The model of the noisy pendulum is used for the description and explanation of many phenomena. Some of these applications have been described in Sec. 1.1 (Brownian motion in a periodic potential, Josephson junction and fluxon motion in superconductors, charge density waves, laser gyroscope, synchronization phenomena, parametric resonance in anisotropic systems, Frenkel-Kontorova model in semiconductors, solitons in optical lattices). For the reader's convenience, we supplement this list by a series of additional applications of the pendulum model: chemical reactions [153; 154], biophysics (neural activity [155], intracellular transport [45], oscillations in the visual cortex [156], penetration of biological channels by ions

[157]), superionic conductors [158], plasma physics [159], surface diffusion [160], electrophoresis [161], rotation of molecules in solids [162], dielectric relaxation [163], polymer dynamics [164], engineering (ship dynamics [165], gravitational gradient pendulum [166]), matter-antimatter asymmetry in the universe [167]. The above list is only a small part of the hundreds of articles describing different applications of the pendulum model. The well-known successful example of ignoring navigation devices, based on the pendulum model, was the voyage of Christopher Columbus which resulted in the discovery of America. At the same time, many positive changes in our life are due to the technological applications of the deep understanding of pendulum dynamics. This subject still attracts great interest.

Bibliography

[1] Wikipedia, the free encyclopaedia.

[2] C. Gauld, Science and Education **13**, 811 (2004).

[3] www.arts.unsw.edu.au/pendulum.

[4] G. L. Baker and J. A. Blackburn, *The Pendulum* [Oxford University Press, 2005].

[5] M. Gitterman, *The Noisy Oscillator* [World Scientific, 2005].

[6] E. I. Butikov, Eur. J. Phys. **20**, 429 (1999).

[7] L. Reichl, *The Transition to Chaos* [Springer, 1992].

[8] I. Lira, Eur. J. Phys. **28**, 289 (2007).

[9] E. T. Whittaker and G. N. Watson, *A Course of Modern Analysis* [Cambridge University Press, 1927].

[10] R. R. Parwani, Eur. J. Phys. **25**, 37 (2004).

[11] A. H. Nayfeh and D. T. Mook, *Nonlinear Oscillations* [Wiley, 1979].

[12] G. L. Baker and J. P. Gollub, *Chaotic Dynamics, an Introduction* [Cambridge University Press, 1990].

[13] D. Permann and I. Hamilton, Am. J. Phys. **60**, 442 (1992).

[14] H. Risken, *The Fokker-Planck Equation* [Springer, 1996].

[15] B. Ya. Shapiro, M. Gitterman, I. Dayan, and G. H. Weiss, Phys. Rev. B **46**, 8416 (1992).

[16] R. Adler, Proc. IREE **34**, 351 (1946).

[17] A. Pikovsky, M. Rosenblum, and J. Kurtis, *Synchronization - A Universal Concept in Nonlinear Science* [Springer, 2002].

[18] M. Gitterman, Phys. Rev. A **35**, 41 (1987); M. Gitterman and D. Pfeffer, Synthetic Metals **18**, 759 (1987); ibid., J. Magn. Magn. Materials **68**, 243 (1987); M. Gitterman, Thermodyn. Acta **169**, 47 (1989).

[19] O. M. Brown and Y. S. Kivshar, Phys. Rep. **306**, 1 (1998).

[20] Y. V. Kartashov and L. Torner, Opt. Lett. **29**, 1102 (2004).

[21] V. E. Shapiro and V. M. Loginov, Physica A **91**, 563 (1978).

[22] N. G. van Kampen, *Stochastic Processes in Physics and Chemistry* [North-Holland, 1992].

[23] M. Gitterman, R. I. Shrager, and G. H. Weiss, Phys. Lett. A **142**, 84 (1989).

[24] J. Kim, A. Sosso, and A. F. Clark, J. Appl. Phys. **83**, 3225 (1998).

[25] R. L. Stratonovich, Radiotech. Electronic (in Russian) **3**, 397 (1958); R. L. Stratonovich, *Topics in the Theory of Random Noise* [Gordon and Breach, 1967].

[26] P. Reimann, C. Van den Broek, H. J. Linke, P. Hanggi, M. Rubi, and A. Perez-Madrid, Phys. Rev. E **65**, 031104 (2002).

[27] S. Lifson and J. J. Jackson, J. Chem. Phys. **36**, 2410 (1962).

[28] L. S. Gradstein and I. M. Ryzhik, *Tables of Integrals, Series and Products* [Academic, 1994].

[29] M. Gitterman, I. B. Khalfin, and B. Ya. Shapiro, Phys. Lett. A **184**, 339 (1994).

[30] E. Heinsalu, R. Tammelo, and T. Ord, arXiv: cond-mat/0208532v2.

[31] G. W. Gardiner, *Handbook of Stochastic Methods* [Springer, 1997].

[32] V. Berdichevsky and M. Gitterman, Phys. Rev. E **56**, 6340 (1997).

[33] I. I. Fedchenia and N. A. Usova, Z. Phys. B **50**, 263 (1983).

[34] M. Gitterman and V. Berdichevsky, Phys. Rev. E **65**, 011104 (2001).

[35] Yu. M. Ivanchenko and L. A. Zilberman, Zh. Eksp. Teor. Fiz. **55**, 2395 (1968) (Sov. Phys. JETP **28**, 1272 (1969)); V. Ambegaokar and B. I. Halperin, Phys. Rev. Lett. **22**, 1364 (1969).

[36] H. Horsthemke and R. Lefever, *Noise-Induced Phase Transitions* [Springer, 1984].

[37] A. J. R. Madureira, P. Hanggi, V. Buonomano, and W. A. Rodrigues, Jr., Phys. Rev. E **51**, 3849 (1995).

[38] M. Marchi, F. Marchesoni, L. Gammaitoni, E. Manichella-Saetta, and S. Santucci, Phys. Rev. E **54**, 3479 (1996).

[39] S. H. Park, S. Kim, and C. S. Ryu, Phys. Lett. A **225**, 245 (1997).

[40] S. L. Ginzburg and M. A. Pustovoit, Phys. Rev. Lett. **80**, 4840 (1998); ibid., Sov. Phys. JETP **89**, 801 (1999); O. V. Gerashchenko, S. L. Ginzburg and M. A. Pustovoit, Eur. J. Phys. B **15**, 335 (2000); ibid., **19**, 101 (2001).

[41] C. Zhou and C.-H. Lai, Phys. Rev. E **59**, R6243 (1999); ibid., **60**, 3928 (1999).

[42] M. M. Millonas and D. B. Chialvo, Phys. Rev. E **53**, 2239 (1996).

[43] Van den Broeck, J. Stat. Phys. **31**, 467 (1983); J. Luczka, R. Bartussek, and P. Hanggi, Europhys. Lett. **31**, 431 (1995).

[44] C. R. Doering ang J. C. Gadoua, Phys. Rev. Lett.**69**, 2138 (1992).

[45] P. Reimann, Phys. Reports **361**, 57 (2002).

[46] V. Berdichevsky and M. Gitterman, Physica A **249**, 88 (1998).

[47] G. H. Weiss and M. Gitterman, J. Stat. Phys. **70**, 93 (1993).

[48] M. R. Young and S. Songh, Phys. Rev. A **38**, 238 (1988).

[49] M. Gitterman, J. Phys. A **32**, L293 (1999).

[50] C.-J. Wang, S.-B. Chen, and D.-C. Mei, Phys. Lett. **352**, 119 (2006).

[51] J.-H. Li and Z.-Q. Huang, Phys. Rev. E **58**, 139 (1998).

[52] J. Isohatala, K. N. Alexxev, L. T. Kurki, and P. Pietilainen, Phys. Rev. E **71**, 066206 (2005)

[53] L. G. Aslamazov and A. I. Larkin, JETP Lett. **9**, 87 (1969).

[54] D. Reguera, P. Reimann, P. Hanggi, and J. M. Rubi, Europhys. Lett. **57**, 644 (2002).

[55] Hu Gang, A. Daffertshoter, and H. Haken, Phys. Rev. Lett. **76**, 4874 (1996).

[56] M. Ya. Azbel and P. Bak, Phys. Rev. B **30**, 3722 (1984); R. Morano, J. Appl. Phys. **8**, 679 (1990).

[57] B. Ya. Shapiro, M. Gitterman, I. Dayan, and G. H. Weiss, Phys. Rev. B **46**, 8349 (1992).

[58] P. Coullet, J. M. Gilli, M. Monticelli, and N. Vandenberghe, Am. J. Phys. **73**, 1122 (2005).

[59] N. F. Pedersen and O. H. Sorrensen, Am. J. Phys. **45**, 994 (1977).

[60] K. Mallick and P. Marcq, J. Phys. A **37**, 4769 (2004).

[61] K. Mallick and P. Marcq, Phys. Rev. E**66**, 041113 (2002).

[62] S. Chandrasekhar, Rev. Mod. Phys. **15**, 68 (1943).

[63] P. Fulde, L. Petronero, W. R. Schneider, and S. Strassler, Phys. Rev. Lett. **35**, 1776 (1975).

[64] M. Borromeo, G. Costantini, and F. Marchesoni, Phys. Rev. Lett. **82**, 2820 (1999).

[65] K. Lindenberg, J. M. Sancho, A. M. Lacasta, and I. M. Sokolov, Phys. Rev. Lett. **98**, 020602 (2007).

[66] R. Harrish, S. Rajasekar, and K. P. N. Murthy, Phys. Rev. E **65**, 046214 (2002).

[67] C. Gregory, E. Ott, F. Romerias, and J. A. Yorke, Phys. Rev. A **36**, 5365 (1987).

[68] J. A. Blackborn and N. Gronbech-Jensen, Phys. Rev. E **53**, 3068 (1996).

[69] W. C. Kerr, M. B. Williams, A. K. Bishop, K. Fessser, P. S. Lomdahl, and S. F. Trullinger, Zeit. f. Phys. B **59**, 103 (1985).

[70] M. N. Popescu, Y. Braiman, F. Family, and H. G. E. Hentschel, Phys. Rev. E **58**, R4057 (1998).

[71] N. Takimoto and M. Tange, Progr. Theor. Phys. **90**, 817 (1993).

[72] L. Machura, M. Kostur, P. Talkner, J. Luczka, and P. Hanggi, Phys. Rev. Lett. **98**, 040601 (2007).

[73] L. R. Nie and D. C. Mei, Eur. Phys. J. B **58**, 475 (2007).

[74] M. V. Bartucceli, G. Gentile, and K. V. Georgiou, Proc. Roy. Soc A **457**, 3007 (2001).

[75] G. Litak, M. Borowiec, and M. Wiercigroch, arXiv-nlin. CD/0607046 v. 1 (2006).

[76] M. I. Clifford and S. R. Bishop, J. Sound Vibr. **172**, 572 (1994).

[77] S. Lenci, E. Pavlovskaia, G. Rega, and M. Wiercigroch, J. Sound Vibr. **312**, 243 (2008).

[78] H. J. T. Smith and J. A. Blackburn, Phys. Rev. A **40**, 4708 (1989).

[79] B. Wu and J. A. Blackburn, Phys. Rev. A **45**, 7030 (1992).

[80] J. A. Blackburn, S. Wik, Wu Binruo, and H. J. T. Smith, Rev. Sci. Instrum. **60**, 422 (1989).

[81] F-G. Xie and W-M. Zheng, Phys. Rev. E **49**, 1888 (1994).

[82] J. P. van der Weele and E. de Kleine, Physica A **228**, 245 (1996).

[83] N. Minorsky, *Nonlinear Oscillations* [Van Nostrand, 1962].

[84] H. M. Lai, Am. J. Phys. **52**, 219 (1984).

[85] L. Gammaitoni, P. Hanggi, P. Jung, and F. Marchesoni, Rev. Mod. Phys.
      70, 223 (1998).
[86] N. G. Stokes, N. D. Stein, and V. P. E. McClintock, J. Phys. A **26**, L385
      (1993).
[87] V. Berdichevsky and M. Gitterman, Europhys. Lett. **36**, 161 (1996).
[88] V. S. Anishchenko, A. B. Neumann, and M. A. Safonova, J. Stat. Phys.
      70, 183 (1993).
[89] F. Marchesoni, Phys. Lett. A **231**, 61 (1997).
[90] Y. W. Kim and W. Sung, Phys. Rev. E**57**, R 6237 (1998).
[91] R. A. Hopfel, J. Shah, P. A. Wolff, and A. C. Gossard, Phys. Rev. Lett.
      56, 2736 (1986).
[92] R. Eichhorn, P. Reimann, B. Cleuren, and C. Van den Broeck, Chaos **15**,
      026113 (2005).
[93] D. Speer, R. Eichhorn, and P. Reimann, Europhys. Lett. **79**, 10005 (2007).
[94] P. Jung, J. S. Kissner, and P. Hanggi, Phys. Rev. Lett. **76**, 3436 (1996).
[95] J. L. Mateo, Phys. Rev. Lett. **84**, 258 (2000).
[96] F. R. Alatriste and J. L. Mateous, Physica A **384**, 233 (2007).
[97] I. Dayan, M. Gitterman, and G. H. Weiss, Phys. Rev. A **46**, 757 (1992).
[98] G. Sun, N. Dong, G. Mao, J. Chen, W. Xu, Z. Ji, L. Kang, Y. Yu, and
      D. Xing, Phys. Rev. E **75**, 021107 (2007).
[99] M. Gitterman, Eur. J. Phys. **23**, 119 (2002).
[100] R. Cuerno, A. F. Ranada, and J. Ruiz-Lorenzo, Am. J. Phys. **60**, 73 (1992).
[101] M. J. Feigenbaum, J. Stat. Phys. **19**, 25 (1978).
[102] J. P. Eckmann, Rev. Mod. Phys. **53**, 643 (1981).
[103] Y. Pomeau and P. Manneville, Comm. Math. Phys. **74**, 189 (1980); P.
      Manneville and Y. Pomeau, Phys. Lett A **75**, 1 (1980).
[104] B. A. Huberman, J. R. Crutchfield, and N. H. Packard, Appl. Phys. Lett.**37**,
      750 (1980).
[105] W. J. Yeh and Y. H. Kao, Appl. Phys. Lett **42**, 299 (1983).
[106] E. G. Gwinn and R. M. Westervelt, Phys. Rev. Lett. **54**, 1613 (1985).
[107] N. F. Pedersen and A. Davidson, Appl. Phys. Lett. **39**, 830 (1981).
[108] A. H. MacDonald and M. Plischke, Phys. Rev. B **27**, 201 (1983).
[109] D. D'Humueres, M. R. Beasley, B. A. Huberman, and A. Libchaber, Phys.
      Rev. A **26**, 3483 (1982).
[110] T. Kapitaniak, Phys. Lett. A **116**, 251 (1986).
[111] M. F. Weher and W. G. Wolfer, Phys. Rev. A **27**, 2663 (1983).
[112] U. Lepik and H. Hein, J. Sound Vibr. **288**, 275 (2005).
[113] H. Hein and U. Lepik, J. Sound Vibr. **301**, 1040 (2007).
[114] H. Seifert, Phys. Lett. A **98**, 213 (1983).
[115] R. L. Kautz and R. Monaco, J. Appl. Phys. **57**, 875 (1985).
[116] J. B. McLaughlin, J. Stat. Phys. **24**, 375 (1981).
[117] L. D. Landau and E. M. Lifshitz, *Mechanics* [Butterworth-Heinemann,
      1976].
[118] R. W. Leven and B. P. Koch, Phys. Lett. A **86**, 71 (1981).
[119] S-Y. Kim and K. Lee, Phys. Rev. E **53**, 1579 (1996).

[120] J. A. Blackburn, N. Gronbech-Jensen, and H. J. T. Smith, Phys. Rev. Lett. **74**, 908 (1995).

[121] J. A. Blackburn, Proc. Roy. Soc. London A **462**, 1043 (2006).

[122] R. Van Dooren, Chaos, Solitons & Fractals **7**, 77 (1996).

[123] P. J. Bryant, J. Austral. Math. Soc. Ser. B **34**, 153 (1992).

[124] J. M. Schmitt and P. V. Bayly, Nonlinear Dyn. **15**, 1 (1998).

[125] A. H. Nayfeh and B. Balachandran, *Applied Nonlinear Dynamics* [Wiley, 1995].

[126] J. Miles Phys. Lett. A **133**, 295 (1988).

[127] P. J. Bryant and J. W. Miles, J. Austral. Math. Soc. Ser B, **32**, 1 (1990).

[128] J. Jeong and S-Y. Kim, J. Korean Phys. Soc. **35**, 393 (1999).

[129] P. J. Bryant and J. W. Miles, J. Austral. Math Soc. Ser. B **32**, 23 (1990)

[130] M. Inoue and H. Koga, Progr. Theor. Phys. **68**, 2184 (1982).

[131] R. L. Kautz, J. Appl. Phys. **86**, 5794 (1999).

[132] S. Boccatteli, G. Grebogi, Y-L. Lai, H. Manchini, and D. Maza, Phys. Rep. **329**, 1203 (2000).

[133] J. Yang and Z. Jing, Chaos, Solitons & Fractals **35**, 726 (2008).

[134] Y. Braiman and I. Goldhirsch, Phys. Rev. Lett **66**, 2545 (1991).

[135] E. Ben-Jacob, I. Goldhirsch, and Y. Imry, Phys. Rev. Lett. **49**, 1599 (1982).

[136] Z. Abbadi and E. Simiu, Nanotechnology **13**, 153 (2002).

[137] P. S. Landa and P. V. E. MaClintock, J. Phys. A **33**, L433 (2000).

[138] M. Gitterman, J. Phys. A **34**, L355 (2001).

[139] Y. Kim, S. Y. Lee, P. Jung, and F. Marchesoni, Phys. Lett. A **275**, 254 (2000).

[140] E. Butikov, Am. J. Phys. **69**, 755 (2001).

[141] T. Leiber and H. Risken, Phys. Lett. A **129**, 214 (1988).

[142] G. J. VanDalen, Am. J. Phys. **72**, 484 (2004).

[143] H. Yabuno, M. Miura, and N. Aoshima, J. Sound Vibr. **273**, 493 (2004).

[144] B. P. Mann and M. A. Koplov, Nonlinear Dyn. **46**, 4227 (2006).

[145] P. S. Landa and A. A. Zaikin, Zh. Eksp. Teor. Fiz. **111**, 358 (1997) (Sov. Phys. JETP **84**, 197 (1997)).

[146] A. Witt and G. Gorelik, Z. Tech. Fiz. Sowjetunion **3**, 294 (1933).

[147] M. G. Olson, Am. J. Phys. **44**, 1211 (1976).

[148] A. Arinstein and M. Gitterman, Eur. J. Phys. **29**, 385 (2008).

[149] P. Lynch, Int. J. Nonlinear Mech. **37**, 345 (2002).

[150] N. W. McLachlan, *Theory and Applications of Mathieu Functions* [Oxford University Press, 1947].

[151] J. A. Blackburn, H. J. T. Smith, and N. Gronbech-Jensen, Am. J. Phys. **60**, 903 (1992).

[152] D. J. Acheson, Proc. Roy. Soc. London A**448**, 89 (1995).

[153] Y. Kuramoto, *Chemical Oscillations, Waves and Turbulence* [Springer, 1984].

[154] V. Petrov, Q. Ouyand, and H. L. Swinney, Nature **388**, 655 (1997).

[155] C. Kurrer and K. Shulten, Phys. Rev. E **51**, 6213 (1995).

[156] H. Sompolinsky, D. Golomb, and D. Kleinfeld, Phys. Rev. A **43**, 6990 (1991).

[157] A. J. Viterby, *Principles of Coherent Communications* [McGraw-Hill, 1966].

[158] W. Dietrich, P. Fulde, and I. Peschel, Adv. Phys. **29**, 527 (1980).

[159] E. M.Lifshitz and L. P. Pitaevskii,*Physical Kinetics* [Pergamon,1981].

[160] P. Talkner, E. Hershkovitz, E. Pollak, and P. Hanggi, Surf. Sci. **437**, 198 (1999).

[161] C. L. Asbury and G. van den Engh, Biophys. J. **74**, 1024 (1998).

[162] Y. Georgievskii and A. S. Burstein, J. Chem. Phys. **100**, 7319 (1994).

[163] J. R. McConnel,*Rotational Brownian Motion and Dielectric Theory* [Academic, 1980].

[164] G. I. Nixon and G. W. Slater, Phys. Rev. E **53**, 4969 (1996).

[165] K. J. Spyrou, Phil. Trans. Roy. Soc. Ser. A **358**, 1733 (2000).

[166] S. W. Ziegler and M. P. Cartmell, J. Spacecraft Rockets **38**, 904 (2001).

[167] M. E. Shaposhnikov, Contemp. Phys. **39**, 177 (1998).

Index